Praise for *The Map of Knowledge*

"Violet Moller brings to life the ways in which knowledge reached us from antiquity to the present day in a book that is as delightful as it is readable."

—Peter Frankopan, author of *The Silk Roads*

"After the fall of Rome, the libraries of the West were burned by marauding Goths and Huns, and the Greek and Roman classics survived only in the Islamic world. Violet Moller's wonderful *The Map of Knowledge* . . . tells the story of how that knowledge was first preserved, then returned to Europe . . . a beautifully written and researched work of intellectual archaeology."

—William Dalrymple, *Spectator* 'Books of the year'

"A sumptuous, glittering, endlessly fascinating book, written with passion, verve and humour."

—Catherine Nixey, author of *The Darkening Age*

"Superb . . . Ambitious but concise, deeply researched but elegantly written, and very entertaining, *The Map of Knowledge* is popular intellectual history at its best."

—*Daily Telegraph*

"An endlessly fascinating book, rich in detail, capacious and humane in vision."

—Stephen Greenblatt, author of *The Swerve: How the World Became Modern*

"As the historian Violet Moller reveals in her expansive book, the passage of ideas from antiquity through the Middle Ages and beyond was fraught with obstacles . . . The story she tells is a fascinating one."

—Daisy Dunn, *Sunday Times*

ALSO BY VIOLET MOLLER

The Map of Knowledge

INSIDE THE STARGAZER'S PALACE

THE TRANSFORMATION OF SCIENCE IN 16TH-CENTURY EUROPE

VIOLET MOLLER

PEGASUS BOOKS
NEW YORK LONDON

INSIDE THE STARGAZER'S PALACE

Pegasus Books, Ltd.
148 West 37th Street, 13th Floor
New York, NY 10018

ISBN: 978-1-63936-837-2

10 9 8 7 6 5 4 3 2 1

Printed in the United States of America
Distributed by Simon & Schuster
www.pegasusbooks.com

CONTENTS

PROLOGUE

BEFORE

The heavens themselves, the planets and this centre
Observe degree, priority, and place,
Insisture, course, proportion, season, form,
Office, and custom, in all line of order:
And therefore is the glorious planet Sol
In noble eminence enthroned and sphered
Amidst the other; whose med'cinable eye
Corrects the ill aspects of planets evil,
And posts, like the commandment of a King,
Sans check, to good and bad. But when the planets
In evil mixture to disorder wander,
What plagues and what portents! What mutiny!
What raging of the sea! Shaking of the earth!
Commotion in the winds! Frights, changes, horrors,
Divert and crack, rend and deracinate
The unity and married calm of states
Quite from their fixture!

William Shakespeare, *Troilus and Cressida*, Act I,
Scene III, 84–101

Mortlake, 1592.

An old man sits before a table, its surface covered with papers. These documents are the record of his life, his achievements and disappointments. They map the many connections he cultivated during his long career: a vast, sprawling network of noble lords, kings and scholars from across Europe whose intellectual interests brought them into the orbit of this remarkable figure. This is his final supplication to the queen he has served for over four decades, herself now an elderly woman too. Evidence of missed opportunities, broken promises and unpaid commissions spill across the table, each carefully listed in the handwritten document he holds in his elegant fingers. Beneath his neat black cap, his eyes are dark and sorrowful, but his beard is long and fine, as white as milk. There was a time, many years before, when his life brimmed with possibility and the queen called him her 'philosopher'. Now here he sits, surrounded by the ghosts of his extraordinary library, his laboratories, his instruments and his patrons, England's great polymath – penniless and reduced to begging.

Dr John Dee is a central figure in the story of sixteenth-century scientific culture. Through his wide-ranging interests, he embodies its scope, its mutability and its strangeness, while the centre of knowledge he created at his house in Mortlake was emblematic of the new types of scientific space that defined that period and influenced the following one. It is said that Dee was in William Shakespeare's head when he dreamt up the character of Prospero for *The Tempest*. His career is certainly representative of the time. Freelance scholars like Dee spent their professional lives navigating complex patronage systems without the safety net of formal roles or regular incomes, pioneers in a burgeoning Protestant intellectual sphere that spread through northern Europe. Often facing the headwinds of religious change, they had to adapt and sometimes even migrate in order to survive and play their part in the new world of learning. We will journey through this world, visiting six places where people gazed at the stars and studied nature, beginning in Nuremberg, home to the first observatory north of the Alps. We will follow in Dee's footsteps from Mortlake to Louvain, Kassel and Prague, and his letters to the Danish island of Hven – the major centres of scientific investigation at that time. Our final destination is another island, described by Francis Bacon in his novel, *New Atlantis* – an imaginary successor to the afore-mentioned centres and the blueprint for modern scientific institutions.

The story of the 'birth' of modern science usually starts with Francis Bacon and René Descartes – credited with establishing natural observation, measurement and experimentation as the cornerstones of the scientific process. But their ideas didn't fall out of the sky. They were forged in the intellectual world of their predecessors. Establishing new ways of doing things often involves denigrating what has gone before and presenting your ideas as novel; Descartes and Bacon were no exception. Their promotion of empirical observation rather than relying on existing texts was not, in fact, a new concept – it had been fermenting throughout the sixteenth century. One of its most radical proponents was the magnificently named Theophrastus Bombastus von Hohenheim, known as Paracelsus, who insisted that those who were genuinely interested in learning, 'must turn to reading not the books of men but the larger book of nature'.[1] This view was shared by many of the people whose stories we will follow; it guided their work and profoundly altered knowledge in countless fields, from medicine to astronomy, alchemy, mathematics and geography.

Just as discoveries are valued above all else in science, histo-rians of the subject have traditionally focused on the 'eureka moments' and the great minds behind them. The places and conditions in which scientific progress was made is a relatively new subject in the history of science, partly because the evidence is thinner, and any reconstructions require a broad, contextual approach. The result is more complex, more nuanced, and ultimately, more interesting. Looking beyond the achievements of Newton and Boyle, of Bacon and Descartes – before the foundation of the Royal Society – to the preceding period, it is clear that without the decades of observational data and discussion about the universe and nature, the technological innovations in printing scientific books and making instruments, to say nothing of the development of dedicated places for studying nature, these heroes of science would have inhabited a very different world.

This is the story of how that world was created during the sixteenth century, specifically how investigation into the natural realm proliferated and flourished in the northern parts of Europe. It is not a well-known narrative, at least compared to the twin megaliths of the Italian Renaissance and the Scientific Revolution, which it falls between and inextricably links. Histories of this period rightly celebrate the achievements of the likes of Copernicus and Galileo, but this narrative is based around places, so these individuals are not my focus. Johannes Kepler is included, because of the role he played in Prague – one of the key cities we will visit. It is impossible to avoid 'great men and big ideas' when writing about the history of science, but I am keen to draw a broader picture by focusing on the figures that sit at the periphery, the places where they worked and the wider culture in which they operated. Where possible, and the sources are very limited, this will include the people, many of them women, in the background who assisted and worked alongside them.

Few of the characters in this story fall into the 'great men' category, and many have been largely forgotten. This needs to be redressed because, as we shall see, they played a vital role in the development of scientific culture in Britain, Scandinavia, Germany, the Low Countries and the Czech lands, and, in turn, scientific culture more generally. Many of their innovations are still integral to how we study the natural world today. The practical developments in precision, communication and technology happened alongside and in tandem with commercial expansion, and in particular, mining, which provided the financial basis for culture across the spectrum, as well as the raw materials – metals – used in the creation of new technology and equipment. This is a tale of commercial success, technical innovation and hard work, of determination, communication and perseverance, of imagination, ambition and triumph, a tale that cries out to be told.

To our rational, orderly, twenty-first-century minds the sixteenth-century map of knowledge appears messy, a paradoxical and confusing place where magic was studied alongside geometry, people searched obsessively for the philosopher's stone and astrology was fundamental to many areas of life. It was also a period of great intellectual change. Knowledge was transformed by the discovery of new lands, new technologies and new visions of nature. Scholars across Europe and beyond shared their ideas with one another, and uncovered ancient texts that brought an array of often contradictory philosophical traditions and theories to light. This makes it slippery and hard to pin down, but all the more compelling for that. We refer to this period as 'early modern', reducing it to a prequel to the main 'modern' event, but the people who lived through the sixteenth century would have been more likely to call it 'the new age', in recognition of the novelty that surrounded them, the inventions, the continents, the ideas and faiths that were appearing on all sides.[2]

By 1500, humanist scholars had already brought large numbers of previously unknown manuscripts to light, texts by Archimedes, Aristotle, Plato, Cicero and Lucretius among a host of others. New, improved translations of known works supplemented these. The idea that ancient knowledge held the deepest secrets of the world was powerful, causing a huge rise in the popularity of occult philosophy, fuelling interest in alchemy and natural magic. One of the most persuasive factors was a collection of manuscripts that purported to contain the writings (in Greek) of Hermes Trismegistus, an ancient Egyptian priest about whom it has been said, 'No ancient writer had an afterlife more active, more paradoxical, or more crammed with incident.'[3] In the early seventeenth century, the Hermetic Corpus was shown to date from the early Christian era, and not around the time of Moses as previously thought – debunking the theory that it contained divine revelation. However, its claim that humans were 'little gods' who could affect nature by using magic remained influential. It runs through attitudes to scientific investigation throughout the sixteenth century and beyond to Francis Bacon, who took it even further.

Viewing the intellectual landscape of this period, the most striking aspect is its variety. In the absence of formal, commonly accepted methods of accumulating knowledge or defined bodies of fact, scholars were open to a wider range of possibilities and more willing to consider conflicting ideas concurrently. We can see this in their acceptance of Aristotelian and Platonic philosophies, their willingness to contemplate both the Ptolemaic and Copernican universal systems, and their enthusiasm for ancient magic alongside empirical observation. If we are to fully appreciate their mindset, we, too, need to embrace uncertainty, to be inclusive and open-minded, and give all areas of knowledge due consideration regardless of their eventual place in the pantheon. We also need to recognise that knowledge was configured differently, that subjects which are unconnected today – astrology and medicine, for example – were once integral to one another.

The stars, or rather astronomy, will be our guide. This was the most prestigious of the mathematical disciplines, one that long played a leading role in the development of science in part because it was often the starting point for investigations of the natural world. People have always built places to observe, to enhance their appreciation and understanding of the night sky. Stone circles, ancient temples, Egyptian pyramids, Gothic cathedrals – all of these are types of observatories designed to frame the movements of celestial bodies at certain times of the year. Astronomers, who devoted themselves to mapping the stars and planets, and interpreting their influence, have played a vital role in almost every civilisation the world has seen. They helped leaders feel confident when making decisions, and ascertained the dates and times which were so crucial to religious practices. They gave people a much-needed feeling of security: reassurance that whatever happened to you in life was meant to be – it was written in the stars. Astronomy naturally led scholars to other areas of knowledge: mathematics, which was needed to plot, measure and predict celestial movements; navigation, which was made possible by knowledge of the night sky; and determining time and the calendar.

In the second century CE, Claudius Ptolemy created the system of the universe that was commonly accepted in Europe and the Arab world for 1,500 years. He presented it as a series of complex mathematical models, enabling scholars to predict the movements of the heavens with a reasonable degree of accuracy. The earth was at the centre with the sun and the planets circling it in a series of large circuits – the geocentric system. But there were clear irregularities in his observations, so Ptolemy made adjustments. He put the planets onto smaller spheres called epicycles, on which they rotated, at the same time as they moved around a much larger circle, the deferent – a bit like couples whirling around the Blackpool Ballroom. He hoped to explain why the planets appear to speed up, change size and go back on themselves sometimes, but because there were still obvious discrepancies, he decided to move the earth away from the centre of the deferent circle and add an arbitrary point, which he called the equant, as the centre of the planets' orbits.

This system was manifestly flawed, but it was a good first attempt, expounded in his vast astronomical masterwork, the *Almagest*. Yet Ptolemy made no attempt to describe or explain the *physical* makeup of the universe – this was the sole province of natural philosophers, right up to the sixteenth century. Natural philosophy, the study of the natural world, was dominated by the works of Aristotle. Aristotle's theory had been that the universe was made up of a series of crystal spheres nested inside one another like Russian dolls, on which the planets revolved. This was a beguiling idea, but it didn't fit with Ptolemy's system. Aristotle had insisted the planets moved at a uniform speed, but Ptolemy could see with his own eyes that they didn't, so he had introduced the equant point to try to solve the problem. However, the equant meant that the earth was no longer at the centre of the universe – another direct contradiction of Aristotle. These ideas existed side by side and in opposition to each other for centuries; many astronomers found them troublesome and attempted to come up with solutions. Eventually, this would lead to the rejection of the geocentric system.

In the sixteenth century, astronomers began invading the territory of natural philosophy, challenging Aristotle and thinking about the physical makeup of the universe, not just its abstract movements. This was one of the most transformative changes the discipline underwent. It meant that people observing the stars in the seventeen hundreds looked for, and saw, completely different things to those in the late fifteenth century, and not simply because they were now using telescopes. They thought about the structure of the universe and assessed its movements in ways that weren't possible a hundred years earlier, in new types of scientific space that had developed over the preceding decades. In part this was sparked by a series of dramatic celestial events in the final decades of the sixteenth century, which provoked a Europe-wide community of scholars to question and reassess foundational theories about the universe. This was the first time a large, geographically diverse group of people observed natural phenomena, then shared and discussed their findings, communicating via letters which were copied and passed on, and occasionally even published, foreshadowing modern academic journals and the peer review process.

In the Middle Ages, astronomers were mainly focused on producing new versions of star tables, known as *zij* in Arabic. These tables were compiled from the *Almagest* into a short, easy to consult text known as the *Handy Tables* and used to predict heavenly movements. Each new version had to be recalculated to the specific coordinates of the place in which it was being written, so they are stepping stones that take us through the history of astronomy, right up to the present day. Between 800 and 1,500, new astronomical tables were produced in Baghdad, Nishapur, Merv, Maragheh, Samarkand and Toledo, the major centres of celestial study in that period where the great observatories of the Islamic world were built. Within Europe there had been efforts to create similar environments in the late medieval period, but on a far more modest scale. The most significant was presided over by Alfonso the Wise of Castile, in thirteenth-century Toledo, where scholars used their observations to produce a new and influential set of star tables using information left to them by the Muslim scholars who had worked in the city in the previous century when it was part of the Islamic Empire.

In this period, astronomy and astrology were not yet separate disciplines. Measuring the movements of the stars and planets and interpreting their effect on earth were carried out by the same people, in the same places. While long discredited by the scientific community, astrology has never lost its grip on the human psyche and remains for some a persuasive system to this day, so it should not be difficult to understand that it held such power for our sixteenth-century ancestors, living in a world of devastating uncertainty. They encountered astrology in various incarnations. If your parents could afford it, they might commission a horoscope when you were born, to give them an idea of the kind of person you would grow into, and the kind of life you might go on to live. In order to do this effectively, the astrologer needed to know the time of conception as well as the precise moment you came into the world so that they could note the exact picture of the heavens at these seminal points in your existence. They would then calculate the relative positions of the houses of the zodiac, noting the particular influences that would determine your fate. Saturn, for example, was cold and dry, heralding a melancholic personality and a tendency to coughing. A person's life was determined by the stars, as was their death, and everything in between. The heavens were mapped onto the body in the grand scheme of correspondences that linked every-thing in Creation into a giant chain of being, with God at the very top. If He was going to send messages to earth, where else would they be written but the sky?

Astrology was an essential part of medicine, taught at universi-ties and used by doctors to diagnose and treat patients every day of their professional lives. It was also used to predict future life events and answer all manner of questions – Where is my missing necklace? Will I get pregnant? Should I take this job? When will the world end? Simon Forman, the leading astrologer of his day, carried out in the region of 8,000 consultations in 1590s London – the Elizabethan equivalent of visiting your therapist. This involved casting a horoscope for the moment and place that the question was asked, or the person fell ill, and mapping the celestial bodies in relation to the twelve astrological houses and signs of the zodiac to determine the influences in play and the effect they would have. Intricate and beautiful, these diagrams required specialist knowledge, something many practitioners did not have; this brought the discipline into disrepute and provoked criticism. While the system was sophisticated and coherent, there had always been flaws. As St Augustine noted, twins would have almost identical horoscopes but could have vastly different lives. 'If one twin is born so soon after the other one that the horoscope stays the same, I expect them to be on a par in everything – which can never be found in twins.'[4] There were other issues: one was how to reconcile the role of free will with the idea that the future was mapped out in advance. Criticism and calls for the reform of astrology gathered momentum during the sixteenth century, amplified by discoveries about the structure of the universe. This reform was a major impetus behind the surge in astronomical observation and accompanying interest among scholars.

Astrology was also considered important for the study of alchemy in certain intellectual traditions, which viewed them as the celestial and tellurian mirrors of one another. John Dee was a leading proponent of this idea, which he discussed at length in the somewhat impenetrable *Monas Hieroglyphica* (1564). For others, however, the connections were superficial, limited to shared names (gold was sol, silver was luna, copper Venus and so on) or the occasional recommendation that an experiment be carried out at a certain point in the celestial calendar. Alchemy was 'an artisanal pursuit concerned with the technologies of minerals and metals',[5] dominated by the quest to transmute base metal into gold using the fabled philosopher's stone. As this dream faded, the discipline gradually transformed into what we now know as chemistry, losing its mystical overtones and becoming fully 'scientific'. Transmutation aside, alchemy was a worthwhile pursuit in so far as it improved methods of extracting and processing metals in mining, with extremely lucrative results. It also involved producing pigments, dyes and artificial gemstones, improving glass and ceramic-making processes, refining salts, distilling perfumes and medicines, and a wide range of other practical concerns. As with so many areas of sixteenth-century knowledge, there were various conflicting interpretations and schools of thought within these practices, leaving plenty of room for contradiction and confusion.

In this period, specialisation did not exist; people did not restrict themselves to a tiny area of study as many academics do today. The stars, the weather, magic, geometry, tides, rhinoceroses, pigments, sacred languages, all were for the taking, all studied in the same place, a place that was often a hybrid museum, library, laboratory and observatory, often inside the family home – the natural world was investigated in kitchens, cellars, attics, bedrooms and garden sheds. It wasn't for the faint hearted. Alchemy was a furious, stinking occupation, fraught with the danger of fire and explosion. Medicine was, quite literally, life-threatening. Patients were usually treated in their own homes because separate, clinical spaces like hospitals were unusual. As studying nature by observing it first-hand grew in popularity, fields, gardens, rooftops and woods became common sites of enquiry too. Astronomers spent their nights out on the lawn or going 'up into the leads, there to consider...the diversities, courses, motions and operations of the stars and planets',[6] necks cricked and breath condensing in the freezing air. Some were employed by kings and princes, provided with a stipend, bed and board, and a place to practise their art; others earned money as doctors, instrument makers or tutors. Almost all of them cast horoscopes for clients – the most lucrative outlet for their astral knowledge. They struggled to make their way at a time when scholarship was informal, unregulated and often unpaid. As yet, there were no institutions with long-term funding, independent of a particular patron whose death would spell the end of the enterprise. 'For a relatively brief time in the sixteenth and seventeenth centuries, the household bridged the gap between the monastery and the laboratory as a site for the practice of natural philosophy.'[7] This 'gap', its peculiarities, its conditions and its inhabitants, will be my focus.

Even after a pandemic when working from home became the norm, it is difficult to imagine contemporary scientific practice happening in a domestic setting. The idea of cooking supper in a laboratory once all the test tubes have been tidied away is explored by Bonnie Garmus in her novel *Lessons in Chemistry* (2022), but the prospect of a cosy home life at the Oxford Science Park is unthinkable (although they do have an onsite nursery and a netball club). In the modern world, science happens in designated places, protected by layers of security, health and safety, and in highly controlled environments. Today, each discipline has its own space, separate and isolated from others. But this was not the case in the early modern period. That separation has happened in the years since – as fields of knowledge have deepened, the places in which they are studied have become specialised, different, siloed off from one another. Today a library is almost always just a collection of books, a laboratory is where scientists carry out experiments wearing lab coats and using test tubes, and an observatory is usually a building housing a giant telescope fixed on the sky. In the early modern period, these sites were diverse, multifunctional and rarely purpose built. Everyone had at least a small collection of books to support their investigations, most had a few items acquired because of their sense of wonder, some had vast palaces filled with exotic animals, strange minerals, and freaks of nature, and many had instruments to aid their work. Observatories evolved to include libraries, laboratories and even printing presses – complete centres of knowledge production that played a pivotal role in the history of ideas. In some cases, special buildings were created to house these centres of astronomical study, while in others the instruments were used and stored in a variety of places: libraries, studies, balconies.

This book will trace the development of these sites, looking at how they were designed and used, what other kinds of paraphernalia they would have contained, who worked in them and how. It will highlight the interplay between domestic and professional spaces – there was no clear dividing line between the two in this period, when only the very wealthy would have enjoyed sleeping in a separate bedchamber, and most people lived in just one or two rooms. Astronomers usually lived in or around the observatory they worked in – they needed to be ready to gaze at the sky when the night was a clear one – and early modern depictions of studies show a book-strewn desk with instruments ranged around it, a rumpled bed in the corner of the room. Wives and children often helped out as laboratory assistants; after all, there is not so much difference between cooking dinner and carrying out chemical experiments. This feels so foreign to us today, but remained common well into the nineteenth century, long after official scientific academies had been founded and science had begun to move into a separate, professionalised space.

The development of the laboratory is a defining feature of early modern alchemy, one which would have a profound influence on the discipline's later transformation into 'Chymistry'. Paul van der Doort's splendid vision of a noble alchemist's laboratory has a high beamed ceiling, an ornate stone hearth with a smoke hood to draw up poisonous gases, closets full of bespoke glass vessels for distillation and portable copper furnaces – the very latest designs money could buy. In contrast, Pieter Bruegel the Elder's 1558 engraving reveals the chaos and despair of a poor alchemist's workshop, smoke billowing and the floor littered with smashed vessels and spilled chemicals. The children climb into the cupboard in a fruitless search for food, and through the glassless window, we see the family's future – eviction and homelessness: a wretched warning about the dangers of becoming obsessed with the philosopher's stone. The two images illustrate the move from alchemist's workshop to chemist's laboratory that was under way. The workshop remained an important site of production for many chemical substances, but the experimental aspects of the discipline would eventually disappear behind the closed doors of the laboratory. The idea that base metal could be transmuted into gold or silver through the power of the mythical philosopher's stone, which could also produce the elixir of eternal youth, continued through the sixteenth and seventeenth centuries – something that seems incredible today. To understand why, we need to imagine a world of much greater uncertainty, a world where once the sun had slipped behind the horizon, only candles flickered in the darkness, marvels appeared readily in the place where legerdemain and credulity merged, where spirits lurked behind the door and almost anything was possible.

Astronomy, on the other hand, had to take place outside, in the dark. The earliest stargazers simply used their eyes to observe the night sky, but as time went by, they began to design instruments to improve precision. In the sixth century BCE, scholars developed an upright rule to measure the length of shadows; other devices followed, including one to measure the diameter of the sun, which was probably invented by Archimedes. By the second century CE, Ptolemy had an array of instruments at his fingertips, simple ones for measurement like quadrants but also more complex astrolabes and armillary spheres which could calculate and predict celestial activity. There are reports in classical literature of planetaria, complete machines which simulated the movements of the heavens, powered by water or air. In 1900 divers discovered something that gives us a glimpse of even greater technical sophistication. While investigating a shipwreck near the Aegean island of Antikythera, they found a machine of breathtaking complexity, made up of intricate interlocking cogs, engraved dials and no fewer than thirty-seven gear wheels that could be programmed to show celestial time, solar and lunar eclipses and a calendar, suggesting that ancient technology was far more advanced than previously believed.

Of all the astronomical instruments developed before the invention of the telescope, clocks were the most significant. Being able to accurately measure time had a profound influence on so many aspects of life, and a singular effect on the accuracy and potential use of astronomical observations. There had been clocks of various kinds for centuries; water clocks were popular in the Arab world and famously reached Europe when the caliph Harun al-Rashid sent one to the emperor Charlemagne – a classic example of one-upmanship masquerading as generosity. In the later medieval period, clockmaking centred on Germany where increasingly imaginative machines with moving figures, bells and whistles were constructed. The most useful in a scientific context, however, were the small domestic devices that could be used alongside other instruments, and these became increasingly accurate during the sixteenth century. All this equipment needed somewhere to be kept and used safely. Some things, notably the mural quadrants, were very large and fixed in situ, making the establishment of observatories, dedicated places for the study of astronomy, essential, besides which observations had to be made from the same geographical location for long periods of time in order to be useful – you can see the heavens from almost anywhere on the face of the earth, but if you want to make regular observations to produce accurate data, you need to make them in the same place using the same instruments, for decades.

In my last book, *The Map of Knowledge*, I followed three major scientific texts as they were transmitted and transformed in the Middle Ages, following them on a thousand-year journey through seven cities that ended in 1500. This is where we will begin, taking up where that narrative left off and travelling to seven places north of the Alps where people studied the stars and made instruments in their quest to deepen their understanding of the world around them. These were observatories, in so far as they were places where people looked up to the night sky and noted down what they saw, often using specially designed instruments that were kept in situ. Beginning in Nuremberg, we will travel across Germany, the Low Countries, Britain, Scandinavia and Bohemia to witness the dramatic rise of commerce that helped to fund scientific endeavour, the innovations in technology that made it possible and the surge of intellectual culture of which it was a part. There is no commonly accepted term for this period in history. The 'Northern Renaissance' usually refers to the fine arts and, in any case, this wasn't a rebirth, as in Italy, with its ancient heritage. To my mind, 'transformation' is a better term, acknowledging that this kind of activity had gone on in earlier centuries, when scholars like Roger Bacon and Albertus Magnus wrote about optics and astronomy and explored the possibilities of technology by using astrolabes and other measuring devices. Like all the other scholars we will meet, they contributed to the rich, convoluted, perpetual story of scientific development, only in different ways to the likes of Copernicus and Newton.

When it comes to scientific culture and structures, the sixteenth century was formative in northern Europe, which has been a centre of scientific innovation ever since. Today, Denmark is home to leading pharmaceutical companies like Novo Nordisk, at Durham University astronomers are using the largest ever supercomputer simulation to map the dark universe and Germany still has an unparalleled reputation for technological innovation and excellence. In 1500 this was just beginning, but over the following decades a dramatic transformation took place. This is the story of how it happened.

There were many key events that set the course for this transformation. Sultan Mehmet II's conquest of Constantinople in 1453, Ferdinand and Isabella's expulsion of Muslim and Jewish communities from Iberia in 1492, Henry VII of England's victory at the Battle of Bosworth Field and Maximilian I's rise to power all played their part. But when it comes to the advancement of learning in northern Europe, three dates stand head and shoulders above the rest: 1450, when Johannes Gutenberg began printing words on paper using a modified cider press; 1492, as Christopher Columbus sailed into the Bahamas archipelago; and 1517, when Martin Luther pinned his *Ninety-Five Theses* onto a church door in Wittenberg. These three phenomena – the printing press, voyages of discovery and the Reformation – created the conditions for the economic, social and intellectual transformations that helped create the modern world.

1

NUREMBERG

CITY OF INVENTION

Our story begins in 1471, in the German city of Nuremberg, several years before Christopher Columbus began planning his voyages, and over a decade before Martin Luther was born. Printing, however, was spreading rapidly through the states neighbouring Mainz, where Johannes Gutenberg had printed his first pages thirty years before. When we hear the name Nuremberg today, we cannot help but think of the terrible events of the last century, first Hitler's rallies and later, the Nazi war crime trials. The city's earlier history and its golden age in the late medieval and early modern period have been obscured. Fifteenth-century Nuremberg was a free city (run by an internal government rather than a ruling family) in the Holy Roman Empire, a vigorous centre of trade home to 40,000 people, saffron merchants and clockmakers among them, the cradle of a new German civilisation. The region was waking up to technology's potential to harness natural resources and generate wealth, something that would become an engine of the capitalist system and the foundation of the scientific enterprise.

As the fifteenth century wore on, Nuremberg became the nucleus of technology in Europe, where instruments were designed and made by a growing, skilled community of crafts-men, the Silicon Valley of its time. Unsurprisingly, printing presses were established there early on and in 1492 the early modern blockbuster, the *Nuremberg Chronicle*, a 'universal history of the Christian world', was published in Latin.[1] The following year, a German edition was in the bookshops, making it accessible to a much wider audience and presaging the huge growth of books in vernacular languages in the following century. Financed by local merchants, illustrated by local artists and produced by local printers, it was a testament to the talent and ambition of the city, one that spread its fame far and wide. Today it survives in greater numbers than almost any other book of the same period.

Like the Fuggers of Augsburg, Nuremberg merchants estab-lished mercantile companies based around their families. Unlike the Fuggers, they were known for being cautious, preferring slow, steady increases in profits to high-risk ventures. Each favoured certain commodities: the Behaim and Ebner families dealt in spices, the Halbachs were wine merchants, while the Landauers focused on copper. They became expert in those markets, eschew-ing moneylending, entrepreneurship and the more adventurous avenues of capitalism that were opening up at that time. Ruling the city through the Large and the Little Councils, they negotiated trade agreements and reciprocal privileges with other cities and regions, stimulating domestic craft industries like bell-making and weapon manufacture by giving artisans access to a wide array of materials and markets across the continent in which to sell their items. As city dwellers free from the constraints of rural life like having to grow their own food, these makers could specialise and focus on perfecting the creation of particular things: clocks, scythes, mirrors, bells, parchment, dice, saddles. City records list

141 separate crafts in 1400, and there were certainly even more by 1500.

In 1568, a Nuremberg shoemaker called Hans Sachs published *The Book of Trades*, in which he wrote short verses describing 'One hundred persons and fourteen, In jobs, professions, Church and State,' each with a woodcut illustration by Jost Amman.[2] The astronomer is depicted seated beside a large terrestrial globe, celestial globes on the table and at his feet. A stone arch opens the room to the sky, and the astronomer is measuring his globe with a pair of compasses. Underneath, Sachs writes: 'The astronomer predicts eclipses and tells by the stars whether the year will be fruitful or one of dearth, war and disease.'[3] There is no instrument maker, but there is a clockmaker, a metalworker and a toolmaker – these were the trades Nuremberg became most famous for. In the early years of the sixteenth century Peter Henlein, a local artisan, made a small portable clock designed to be worn around the neck or fastened onto clothing – the first known watch, called a 'living egg' because of its shape and the miniscule steel cogs that turned inside it. Henlein's workshop produced hundreds of these and other innovative timepieces, table clocks, pocket watches, some to hang from coaches, others designed to look like books, and even an alarm clock with a flint to light a candle.

Innovation was a major preoccupation in Nuremberg. Hans Ehemann invented a keyless lock, Georg Hartwig developed a calibre system for guns and Sebald Behaim cast one of the heaviest cannons ever made. The city authorities inspected every item made before it could be exported to check it met their exacting standards of quality. Weights, ornately carved pistols, brass trumpets and compasses were among those receiving the 'N' or eagle stamp to show they had passed the inspectors' scrutiny – this was fundamental to promoting and preserving the city's reputation for integrity. It also attracted the best, most ambitious artisans to come and set up workshops, to take advantage of the commercial opportunities and community of like-minded people.

On 4 July 1471, the German astronomer Johannes Regiomontanus wrote to a friend:

> Quite recently I have made [observations] in the city of Nuremberg…for I have chosen it as my permanent home, not only on account of the availability of instruments, particularly the astronomical instruments on which the entire science of the heavens is based, but also on account of the very great ease of all sorts of communication with learned men living everywhere, since this place is regarded as the centre of Europe because of the journeys of the merchants.[4]

Regiomontanus was one of the first figures, and arguably the most influential, to bring scientific knowledge north from Italy in this period, and yet he is almost unknown outside specialist histories. There is only one biography of him in English (translated from German) and his presence in Nuremberg today is slight, although there is a small observatory named after him. The son of a miller, he was christened Johann Müller, but like so many of his contemporaries, he was known by the town he came from: Königsberg, hence the Latinised nickname Regiomontanus – king's mountain. History does not relate much about his early life, or what the locals made of this extraordinary boy. They didn't have much time to get the measure of him because he left home to matriculate at the University of Leipzig in 1447, aged just eleven, young even for the time. Three years later, having exhausted the educational possibilities on offer in Leipzig, he was on the road to Vienna, where he had heard he could study maths and astronomy under a scholar called Georg Peuerbach. Regiomontanus lost no time in completing his university courses, quickly reaching the level of a master's. However, university regulations meant that he had to wait until he was twenty-one to graduate. It can't have been a surprise to anyone that he elected to stay on to teach and research with Peuerbach. They made observations together and Regiomontanus learned how to construct instruments to improve their accuracy. Both men were alarmed by the discrepancies they found in the *Alfonsine Tables* of Toledo so improving the data became a priority. Then, in 1461, disaster struck – Peuerbach died suddenly, aged just thirty-eight. Regiomontanus had lost his collaborator and friend, at a time when astronomers were thin on the ground.

Fortunately, he had met someone the year before who could help. Cardinal Johannes Bessarion had come to Vienna on a diplomatic mission from Italy to gain support against the Ottomans. Bessarion was a learned, enlightened man, who had been born in Trebizond on the Black Sea coast and brought up in Constantinople in the Orthodox tradition. His lifelong wish was to reunite the two Christian churches, and he worked ceaselessly to bridge the cultures and transmit knowledge from the eastern, Greek-speaking world to the Latin west, bringing hundreds of manuscripts (including a copy of Ptolemy's *Almagest*) to Rome where he made them available to scholars. Regiomontanus must have felt he had entered the very gates of heaven when he arrived at Bessarion's elegant house and saw the library. Regiomontanus taught Bessarion astronomy and maths, receiving tuition in Greek in return, which he then used to make new, improved translations of ancient texts. He constructed a brass astrolabe for the cardinal, inscribing it with the dedication: 'Under the dominion of the divine Cardinal Bessarion I Johannes's work appears in Rome,

1462.'[5] The combination of Greek language with maths and astronomy was central to Regiomontanus' work, and in his view, essential to the progress of those disciplines. He and Bessarion spent the next four years travelling around Italy together and working at the house in Rome, which was a hub for other émigrés from the east and a centre of translation from Greek into Latin at a time when the city was a vibrant centre of scholarship and book collecting.

In 1467 Regiomontanus was tempted back over the Alps by an offer from Matthias Corvinus, king of Hungary, whose recent victory against the Turks had left him in possession of several rare manuscripts. Unable to resist the prospect of discovering new texts, Regiomontanus set off northwards to the Hungarian court in Buda, a rare beacon of humanism outside Italy. A few years later, still in Matthias' service, Regiomontanus wrote a letter to a fellow scholar at the University of Erfurt, keen to recruit others to make observations and also to open up dialogue on astronomical and mathematical problems with the wider community, something he had done while in Italy. This letter, written in July 1471, is as an invaluable source of information on his plans, as it is for revealing the internal state of these areas of knowledge. One of his priorities was to calculate new planetary tables based on his own improved observations; another was to set up a printing press to publish a selection of scientific works. He then listed thirty-eight questions or exercises in astronomy, maths and astrology, some his own creation, others from existing sources. They give us a clear picture of where the limits lay, and what the preoccupations were at the time, but the recipient must have found them overwhelming – he never replied – so Regiomontanus' nascent discussion group didn't get off the ground. At this stage the scientific community in northern Europe was vanishingly small – there were very few people who shared his interests and fewer still who could confer at his level. But that was about to change.

Regiomontanus' time in Italy with Bessarion was formative. It was here he encountered the wonders of ancient and Arabic science through the books collected by his patron and the other scholars he met on his travels. Occasionally, he discovered manuscripts himself, like the copy of Diophantus' *Arithmetica* he found in Venice in 1462. This breadth of knowledge was not available anywhere else in Europe; Italy was the main beneficiary of manuscripts brought from Constantinople after it was taken by the Ottomans in 1453, as well as the focus for the general recovery of ancient texts – especially those written in the original Greek and unmarred by centuries of translation via Arabic and Latin – that was the central focus of Renaissance humanism.*

In 1471, King Matthias sent Regiomontanus to Nuremberg to work on a new set of astronomical tables based on new, improved observations. On 29 November the city council granted him permission to live in the city until the following Christmas. He took on a house and began setting it up as a centre where he could assess, correct and produce information of the highest standard before disseminating it. Once the manuscripts were perfected, they needed to be printed, so Regiomontanus had a press installed. For the first time, the whole process of knowledge production was under one roof, controlled by one person. There were huge

---

* Spain was no longer the diverse place it had once been; from 1492 onwards it would become even less so, as Ferdinand and Isabella imposed their vision of monolithic Catholicism. In 1499 Cardinal Ximénez de Cisneros was sent to the ancient city of Granada to remove any traces of the Muslim and Jewish culture that had flourished there for seven hundred years. He ordered every book in Arabic or Hebrew to be brought to the main square and burnt on an enormous bonfire – a pyre of scientific ideas, literary beauty and countless other wonders – just one episode in the sequence of events that left Spain with only a handful of Arabic manuscripts in the following century.

challenges involved in printing mathematical and astronomical books, which, being full of complex tables of numbers and diagrams, were particularly susceptible to errors, and publishing them himself was apparently the only way Regiomontanus could ensure their accuracy. This was not a minor undertaking. Regiomontanus needed specialist typesetters, draughtsmen and woodcut makers in order to get started – and just two decades after the invention of movable type, people with these skills were not easy to find. The great scientific printing houses lay in the future; Erhard Ratdolt's innovations came after Regiomontanus' death in 1476 and Aldus Manutius did not establish his press in Venice until 1494.

The first book Regiomontanus published was Peuerbach's *New Theory of the Planets* (*Theoricae novae planetarum*, 1472), a tribute to the man who had been such a formative influence and collaborator. It went on to be one of the standard astronomical textbooks of the sixteenth century, pored over by students in towns and cities across Europe. He revealed the other books he intended to publish in a document he called his *Tradelist* (1474), which he sent to several universities. This is the main source of information we have for the enterprise. It is divided into two parts, first works by other authors, second those by himself, 'which, whether they were to be produced or not, innate modesty and the republic of letters long debated amongst themselves. Reason determined they should be attempted.'[6] The list contains several works by Ptolemy, including his masterpiece on astronomy, the *Almagest*, and Euclid's foundational text on mathematics, the *Elements* (neither had been printed with their diagrams before), along with several of Archimedes' most important works, other seminal ancient treatises on mathematical and scientific subjects, and various maps. Several of his own works are commentaries correcting false assumptions and bad translations of the titles in the first list, in addition to original works on, for example, rays, comets, weights and aqueducts, and 'burning mirrors and other things of many kinds and occasioning wonder'.[7]

Regiomontanus was thus preoccupied with addressing errors and establishing a commonly accepted, accurate body of fact on which to base enquiry into the natural world. He compared and collated as many copies and translations of a manuscript he could lay his hands on to produce the best, most accurate, version of the text, processing and correcting the astronomical canon and expressing it with clarity, making it more accessible and easier to build upon. Taken together, these books are foundation stones of modern science. Their influence had already been profound, and would continue to be so, especially after they were reproduced in their thousands by the printing press – not the one Regiomontanus set up, but others, mainly in Venice. By printing this list in Nuremberg and announcing his intention to publish them there, Regiomontanus brought the newly rejuvenated appetite for study of the natural world, initially via texts and observation, to the lands north of the Alps, helping to prepare the ground for the following century. It was a manifesto for how scientific work should be carried out, a template for succeeding generations of scholars. It confirmed Nuremberg's unrivalled position as the northern centre of technology and innovation.

At the end of the *Tradelist*, Regiomontanus talks briefly about the other aspects of his cultural enterprise. In the same workshop, 'There shall be made also astronomical instruments for celestial observations. And also other things for common daily use,'[8] doubtless designed and overseen by the master. The workshop was a major intersection between craft and scholarship. In this period, if you wanted an astrolabe, you either had to make one yourself using a manual or specially commission one from a goldsmith. There were no dedicated instrument shops, but as scholarship spread in Europe, more and more people became interested in measuring the stars, transmuting metals and distilling tinctures. As the demand for astrolabes, glass vessels and other specialist equipment rose, people started making them to sell, setting up centres of production to cater for new markets. Scholars and their patrons also opened workshops where metal could be fashioned into measuring instruments. Regiomontanus was a pioneer in this field, and he likely lived and worked in the same place, as most people did in this period. The *Tradelist* mentions of a planetarium or astronomical clock being made in the workshop, 'a work clearly to be gazed upon as a marvel',[9] and in the treatise he wrote about armillary spheres Regiomontanus refers to a geared clock, which he used to note down the time when taking celestial measurements; living in the European centre of clockmaking was central to his project.

The printing press also required technical equipment, and in particular, the type, made by carving letters into a soft metal and then casting them in a special alloy of lead, tin and antimony, which was durable and well suited to repeated use in the press. This was a very new technology so Regiomontanus would have needed to find someone with printing experience to come in and train others. He also needed skilled woodcut carvers who could produce the elegant diagrams that looped across the pages of books like the *Almagest* and the *Elements*. Printing them was a serious challenge, but without them the text was impossible to understand. It is highly likely that one of Regiomontanus' employees in the press was a young man from Augsburg called Erhard Ratdolt. Ratdolt's father was a woodcarver and town archives show that this is what Ratdolt was doing in Augsburg in the

1460s. Then he disappears from the record, only to resurface a few years later in Venice, where he opened a press with two associates. The very first book they printed was Regiomontanus' *Calendarium*, which could not have been done without a manuscript of the original text, begging the question, how and where did Ratdolt obtain it? The obvious (but unprovable) answer is that he spent time in Nuremberg working at the press with Regiomontanus, who gave him a copy of his work to take to Venice and print. This theory is further strengthened by other books Ratdolt went on to publish, several of which were on Regiomontanus' *Tradelist*.

Dangerous, dirty and difficult, publishing was an exceptionally challenging career at this point, and many early printers went out of business within the first couple of years. It was hard to predict sales and make sound business decisions because the industry was so new; printing houses were a health and safety nightmare – vats of boiling pitch to make ink bubbled near to the presses themselves, which were heavy and difficult to operate. Ratdolt was not fazed by this, however; on the contrary, he was ambitious – some may have said foolhardy. Following Regiomontanus' list, he decided to print Euclid's *Elements*, complete with its myriad diagrams. To achieve this monumental feat, he had to carve 420 separate woodcuts which tumble down the special, extra-wide margins – it was a milestone in printing technology and a tribute to Regiomontanus' legacy, and earned Ratdolt the praise of fellow printers. J. Sandritter wrote, 'The seven arts, abilities granted by the divine power, are amply bestowed on this German from Augsburg, Erhard Ratdolt, who is without peer as a master at composing type and printing books. May he enjoy fame, ever with the favour of the Sister Fates. Many satisfied readers can confirm this wish.'[10]

As master of his own independent, self-sufficient research institution, Regiomontanus was able to impose high standards of intellectual rigour and textual accuracy on the books and information he produced. His model of knowledge production, relatively free from the constraints of royal patrons, university rules and mainstream publishing, inspired scholars in the following centuries, and was emulated several times. In the workshop, local craftsmen made astronomical instruments to Regiomontanus' specifications – a landmark moment in the history of their production and one that helped to cement Nuremberg as a major centre of metal working and instrument making. Accurate, complex measuring devices and calculating machines were becoming increasingly indispensable to the study of astronomy, and this reliance on technology would only grow over the following centuries.

On 28 July 1475, Regiomontanus noted down his last observation. He left for Rome soon after, possibly in response to a summons from Pope Sixtus IV, who wanted to consult him on calendar reform. Discrepancies in the calculation of Easter had been a concern for some time, highlighted by both Bessarion and Regiomontanus, and it was a problem that would rumble on through the following century. He died in Rome the following summer, probably of plague, which ravaged the city that year. Later writers were keen to embellish the threadbare story of his death; one claimed the Pope had made him Bishop of Regensburg, another that he had been buried in the Pantheon. More salaciously, there was a rumour that he had been poisoned by the sons of a scholar whose work he had criticised. There is no evidence to support any of these claims; we are completely in the dark about his last months. He was just forty years old and left a huge amount of unfinished work behind him – only nine titles had been published, and not all of them were on Regiomontanus' original list of forty-seven.

The press was shut down and the project abandoned. But there was a group of like-minded scholars who were determined to continue his legacy, most notable among them Bernhard Walther, originally from Memmlingen, who Regiomontanus taught and worked with in Nuremberg. There is no evidence to support the suggestion that Walther was his financial backer, however, he must have been a person of some means. Regiomontanus had been making celestial observations regularly since his time in Vienna with Peuerbach; these were a fundamental part of his astronomical project and Walter worked as his assistant. Regiomontanus described using a *regula ptolemaei* to observe the altitude of the sun and a lunar eclipse on 2 June 1471, and a Jacob's staff to measure the distances between the stars and the comet of 1472, but we have no idea where he used them or whether he had a special observatory space in his house. Claims that Walther built one for him in a house on Spitzenweg only began in 1800; if this were true, it seems strange there is no earlier mention. If he was only using a couple of instruments and was used to making observations on the hoof, so to speak, in a wide range of places (throughout his travels in Italy, his time in Hungary), perhaps he did not see the need for a dedicated observatory, only a good view of the night sky. As we will see, observatories could range from a storeroom adjoining a balcony to an entire palace with retractable domed roofs and large-scale, built-in equipment. The latter were expensive, long-term projects, far beyond the reach of mendicant scholars like Regiomontanus. The former, however, were more attainable and Walther created exactly this type of set-up when he moved house in 1501. He began making observations on 2 August 1475, just a few days after his master's last entry, for the most part mapping the locations of planets in relation to a selection of the fixed stars. He used these calculations to help accurately remount and calibrate his armillary sphere and *regula ptolemaei* (which was made of brass and 2.5 metres long) on their stone bases at the beginning of each observational session. These instruments produced some incredibly accurate results, especially when it came to eclipses, and Walther was a gifted observer, blessed with a steady hand and excellent eyesight.

When Matthias of Hungary sent an envoy to enquire about buying his old astronomer's books and instruments, they were told that they were not available because Walther already had them; he was using them to continue the observational programme begun by Regiomontanus. Walther later explained, 'As to the majority of my books and all of my instruments, except for the armillary sphere, which have come to me by purchase equally from the heirs of that most ingenious of his art Master Johann Müller of Königsberg, and from Konrad Scherp.'[11] After Walther died in 1504, they were sold off and dispersed, in direct contravention of Walther's plea that they should 'be kept together and not split up or divided up'.[12] Several ended up in the Nuremberg collections, where they remain to this day. Mathematical printing continued, and in the following decades the city became a flourishing centre with Regiomontanus' own *De Triangulis* (1533), Copernicus' *De Revolutionibus* (1543) and Cardano's *Ars Magna* (1545) testament to 'Regiomontanus' importance, not only as a mathematician and astronomer, but also as a publicist and architect of the renaissance of mathematics'.[13]

The house Walther purchased in 1501 was impressive; it still towers over the main square, just below Nuremberg castle. He had two windows and a balcony built onto the top floor of the southern gable and installed his instruments there, creating a modest, yet ground-breaking, observatory – the first identifiable one in northern Europe. The building in question has survived and is now a museum. But it's not dedicated to Walther, Regiomontanus or astronomy. It is celebrated as the home of the artist Albrecht Dürer, who purchased it in 1509.

*   *
*   *
*

The same year that Regiomontanus had arrived in Nuremberg, the Dürer family had welcomed a son. As was traditional, his god-father chose the name: Albrecht, after the baby's father. Albrecht senior, a goldsmith, noted, 'in the sixth hour on St Prudentius's Day, the Tuesday before Ascension Day, my wife Barbara bore me my second son.'[14] The specific time and date would enable astrologers to draw up his nativity. Originally from Hungary, Dürer's father had settled in Nuremberg in 1455, married his master's daughter and built a successful business. He counted the Holy Roman Emperor Frederick III and the Bishop of Poznań among his clients and was made the precious metal tester for the city, in charge of ascertaining value and quality. As with all craft families, Albrecht studied at his father's business and was on the way to becoming a goldsmith too, until, at the age of fifteen, he decided he wanted to be an artist instead and joined the workshop of Michael Wolgemut, the most famous in the city, as an apprentice. He didn't look back.

Exactly three years later Dürer's apprenticeship ended, and he set off on the traditional journeyman years of travel to learn new skills and establish a network. When he returned to marry Agnes Frey, the daughter of a wealthy brass maker, he was a master of copperplate engraving, an almost unknown art in Nuremberg. Soon after the wedding, he abandoned his bride and travelled south through the mountain passes to Venice, a two-week journey he probably took with one of the many merchant convoys who carried goods between the two places. There were many reasons for him to visit the magical city on the lagoon, but high on the list must have been visiting its printing presses and gathering expertise and contacts for the venture he was about to launch in Nuremberg. On his travels, he was inspired by the work of artists like Mantegna and Bellini, and learned about anatomy and perspective from Jacopo de' Barbari, setting him on the path of the mathematical study of art he would follow for the rest of his life. Home again in Nuremberg, he used Agnes' ample dowry of 200 florins to open a workshop.

In 1505 Dürer set off for Venice again, but this time things were different. He arrived as a well-known artist: wealthy, celebrated, his prints selling in local shops, his monogram recognised throughout the watery city. However, he was only known as a draughtsman, not a master of the rich colour palettes that characterised Venetian art, and he was desperate to show his talent. The powerful German merchant guild commissioned him to paint an altarpiece for the nearby church of San Bartolomeo (as we shall see, *The Feast of the Rose Garlands* eventually ended up in Prague, being admired by Rudolf II). Dürer was feted in Venice. Nobles sought him out, and the Doge himself even came to the studio to watch him painting. 'Here I'm a gentleman,' he wrote to his friend Willibald Pirckheimer, 'at home [in Nuremberg] a parasite.'[15] In Venice, artists were worshipped; this was how Dürer wanted to be treated.

By 1509 he was back at home, well established as a formidable talent and man about town, flush with Venetian ducats. It was time to move to larger premises, so he took on Bernhard Walther's enormous house just below the castle and set it up as a major centre of cultural production and innovation. The workshop gave

Dürer control, just as it had Regiomontanus. Here he was able to oversee every stage of his cultural output, from initial design to finished painting or print. Dürer's success in this endeavour, along with the house's preservation, give us unprecedented access to one of the most important and innovative workshops there has ever been. It is a portal into the sixteenth century and the life of Albrecht Dürer.

The artist moved in with his wife Agnes and his mother, Barbara Holper, who ran the house and oversaw the business of selling prints when he was away. The cheaper woodcuts were for the mass market, the finer copperplate engravings for the wealthier customer. The *Rhinoceros*, made without ever beholding the creature itself, was his bestseller, going into eight editions. Customers were encouraged to colour in the prints themselves. Members of the new urban middle class, they were one of the first generations with money to spend on decorating their homes with pictures, luxuries and fripperies with no practical value – pioneers in the burgeoning realm of consumer culture.

Dürer was the first artist to take advantage of the printing press by making multiple copies of his works and selling them on the open market. According to the founding father of art history, Giorgio Vasari, he inspired Raphael to have engravings made of his own works – it was a two-way street between Italy and the north when it came to cultural innovation. There were printing rooms at the house, and other workspaces for the highly skilled woodcarving and etching work. The painting studio was in a large room facing north-east with the best light, where he created portraits of wealthy clients and other masterpieces like *Adam and Eve* and the *Adoration of the Trinity*. There were also dedicated spaces where his assistants ground up precious shells and minerals to make the paint, carefully mixing the powders with linseed or walnut oil to get the right consistency. The house thronged with life: Agnes and Barbara busy with the prints and account books, issuing orders to the household servants and specialist assistants who operated the presses; apprentices clattering up and down-stairs carrying canvases and paints for the master; and dogs, everywhere, getting under your feet – Dürer's feeling for them radiates from every drawing he made. You can tell he knows them and loves them, strokes them, understands their contours, is familiar with the feel of their fur, can read the meaning in their eyes. This love of dogs extended to the entire natural world. Many of the creatures that populate his pictures would have lived in the house in Nuremberg, alongside the non-living natural wonders he owned.

Dürer was awestruck by the natural world, obsessed with stud-ying and capturing it. In 1503, he turned his forensic gaze upon a patch of weeds, dug from the surrounding countryside and carried back to the studio where, using pen, ink and watercolour, he produced an image of ground-breaking naturalism and beauty. Every plant in the *Great Piece of Turf* is identifiable, each blade of grass perfectly rendered. This study of nature is scientific in detail and accuracy. Even the roots and soil are shown; it is the first image of its kind. On journeys Dürer exchanged prints for natural curiosities: an Indian coconut, pieces of coral, bamboo arrows, a large tortoise shell, two parrots in a cage – all packed up and sent back to Nuremberg, to be immortalised on paper or copperplate. But unlike the curiosities amassed by modish aristocrats, impris-oned on velvet cushions in the darkness of ornate drawers, Dürer's marvels lived forever. His work seethes with them, he touched them, he drew them and painted them, examined them obses-sively, he marvelled at them, they populated his dreams and enliv-ened his fantasies.

Dürer was a trailblazer in this new world of artistic power, communication and scrutiny of the natural world. In around 1497, he designed his own monogram, an idea copied from his hero the painter Martin Schongauer. But Dürer took it to another level, inscribing it on most of his works, often obvious, but sometimes hidden within the image: on a plaque in the centre of *Adam and Eve*, on a bookmark in his depiction of *St Jerome*, a stroke of genius that helped build the Dürer brand, an early modern Apple logo. Keen to emulate Regiomontanus' workshop and printing press, and continue his legacy, Dürer's workshop became a source of inspiration in its own right by demonstrating to others later in the century the value of being able to control cultural production. As we shall see, John Dee dreamed of having his own press, and Tycho Brahe built one, along with a paper mill to feed it.

Inspired by what he had seen in Italy and by Regiomontanus' enterprise in Nuremberg,* Dürer began a new cultural movement in Germany that spread far and wide, liberating the imagination, and transforming the way society looked upon the artist. He was a man on the rise, blatant in his desire for money, power and success, keen to make the most of the new opportunities for social mobility – he stares out of his self-portraits, daring us to question his right to advance, to succeed, to be splendid. Gorgeous in his silk robes, he stands at the beginning of a transformative period for the status of the artist, and the scholar. By the beginning of the seventeenth century, both can expect very different lives. They can

---

*   Dürer was a leading member of the circle of intellectuals who saw themselves as Regiomontanus' successors, men like Walther, Willibald Pirckheimer, Johannes Werner and Johann Schöner. The city archives record that on 13 November 1522, 'we sold Albrecht Dürer 10 of Bernard Walther's books for 10 gulden'. Any or all of these books could have formerly belonged to Regiomontanus – we know for sure that Dürer owned one of his copies of Euclid's *Elements*, but it was unfortunately lost in the seventeenth century. (Zinner, *Regiomontanus, His Life and Work*. Oxford: North-Holland, 1990, p. 168.)

rise from the lowly undergrowth of craft to the heady heights of the court – we will follow them on this journey. To achieve this, they will need patrons – generous, enlightened patrons – who will support, guide and nurture them throughout their careers. Previous generations of scholars (and many artists) had been monks, with no need to earn a living from their studies, no commercial angle to their lives aside from the soothing background hum of monastic wealth. Dürer had to seek commissions and livings from merchants, the nobility and, the ultimate source of patronage, the Holy Roman Emperor Maximilian I. The Habsburg dynasty was the foremost funder of culture in this period; its influence on art, technology and the intellectual world was second to none.

Patronage powered culture in this period absolutely. There were no official institutions to provide education and financial support to artists and scientists, no grants to apply for or large companies to step in with sponsorship, so patronage was vital to cultural life. It bound societies together with reciprocal relationships that provided the wealthy party with prestige, support, information and pleasure, while the client or vassal benefitted from money, protection, advancement and contacts. The lives of Maximilian's descendants intertwine with scholars, artists and craftsmen in a web of symbiotic relationships that determined the direction of knowledge production and produced some of the greatest works of art in western Europe. The precedent they set filtered down through society during the sixteenth century, inspiring generations of lesser rulers, aristocrats and wealthy individuals to plough their money into beautiful things and beautiful ideas. In the merchant class, the Fugger family surpassed all others. They were northern Europe's answer to the Medicis and played a pivotal role in the movement of cultural dominance from

Italy to Germany and the Low Countries. Put simply, they were the most important figures in the economic growth of northern Europe, without which none of the art, culture or technology could have happened. To meet them, we need to leave Nuremberg and head south for two days, on the well-trodden road to its main rival, another imperial free city, Augsburg.

*
*   *
*

Founded by the Romans in 15 BCE in the luxuriantly wooded foothills of the Alps, Augsburg flourished as one of the first major cities merchants descended to on their way from Italy, through the Alpine passes of the Via Claudia Augusta. Visiting the area at the end of the sixteenth century, the Englishman Fynes Moryson mulled over the relative merits of the two places:

> The City [Nuremberg] is absolute of it selfe, being one of the free Cities of the Empire, and mee thinks the chief, or at least second to *Augsburg*: surely it may perhaps yield to *Augsburg* in treasure and riches of the City, but it must be preferred for the building, which is all of free stone sixe or seuen roofes high: I speake of the whole City of *Augsburg*, for one street thereof is most beautifull, and some Pallaces there are fit for Princes, of which kind *Nurnberg* hath none.[16]

Like the Medicis in Florence, the Fugger family elevated being extremely wealthy to an art form. The French writer Michel de Montaigne was in Augsburg in 1580 and visited, as many travellers did, the Fugger residences. He wrote afterwards in his diary that 'We were permitted to see two rooms in their palace: one of them large, high and with marble floors, the other one low and filled with old and modern medallions, with a small cabinet at the back. These are the most magnificent rooms I have ever seen.'[17] They may not have had royal blood flowing in their veins, but they lived in splendour and backed every royal house on the continent, loaning sums of money so eye-watering that, by the 1550s, Spanish cargoes of silver were sailing from the Americas straight into their bank accounts, via Antwerp docks.

Even compared to today's super-rich, the Fuggers were exceptional. They had silver mines in Tyrol, lead in Carinthia, copper in Slovakia and offices in Rome, Antwerp, Kraków, Lisbon, Venice and Ofen (Budapest), and agents everywhere from Gdansk to Cadiz. They dominated the mercury market, vital for making mirrors, for medicines, to process gold and in alchemical processes. From 1516 they supplied copper to Pflach to make into brass, helping to increase production of the metal favoured by instrument makers and stargazers to make the astrolabes, clocks and quadrants that transformed our knowledge of the heavens. More than any other family in this period, they were responsible for transforming mining into an industrial enterprise, helping to forge a metal framework that supported northern European expansion.

The Fugger family originally arrived in Augsburg in 1367, settling in to work as weavers at a cloth workshop. Later, they moved into finance, lending money to perennially broke local nobles like Sigismund, Archduke of Tyrol, who repaid them not with interest, but in the rights to purchase silver at low prices direct from the mines, allowing them to sell it on for profit. By 1511 the company was worth about 197,000 florins; sixteen years later, their assets were estimated at three million florins, and by 1536 this had risen to 3.8 million – one million was owed to them in Spain alone. The Fuggers and the Habsburgs were bound together by the rivers of money that flowed between them. Jakob

Fugger, known simply as 'the Rich', first loaned money to Maximilian I in 1491. This initial loan of 120,000 florins brought the firm 30,000 marks of silver that year alone, and it was the first of many. In 1519, when Maximilian's successor Charles V needed money to ensure his election as Holy Roman emperor, Jakob Fugger was only too happy to oblige. During his reign, Charles V borrowed 28 million ducats, thirty-six per cent of which originated in Augsburg – 5.5 million from the Fuggers and 4.2 million from the other great merchant family, the Welsers. The breakdown is a revealing insight into the finances of Europe in the period 1521–55; it shows that the Genoese supplied a similar amount to Augsburg, Antwerp ten per cent and the remainder was sourced in Spain itself – at this point, Germany and Italy were level pegging when it came to imperial finance. The Habsburgs used this money to pay for prosecuting wars, organising coronations, celebrating weddings, for travel, embassies and, of course, art, scholarship and instruments.

In the last decade of the fifteenth century, the Fuggers stormed the greatest institutional bastion of them all, the Catholic Church. They began handling the transfer of *servitia*, the money paid for benefices and ecclesiastical positions, taking over contracts traditionally handled by Florentine bankers. Before long, they were lending money to the Pope as well. They set up an office in Rome, which was run by Jakob's associate, Johann Zink, who was also a cleric, allowing him to collect no less than thirty-two benefices for himself along with curial offices, honours and position as a papal *familiar*. Zink was in charge when the Roman office financed the recruitment of 150 Swiss soldiers in 1506 and paid their first month's salary when they arrived at the Vatican – the Swiss Guard still looks after the Pope today, dressed in the least practical military uniform on earth.

Zink was another Augsburger, as many Fugger employees were. When it came to raising credit to lend, the family usually looked to their own members or local families like the Höchstetters and Welsers, successful business dynasties who collaborated and competed with one another in equal measure. They formed the close Augsburg faction of a Europe-wide network of associates, clients and acquaintances held together by personal relationships, transactions and mutual benefit. Jakob Fugger travelled regularly to visit people face to face and cement the bonds between them, always arriving with splendid presents to bestow. Gift-giving was an integral aspect of society; the Fugger accounts regularly record huge amounts for 'the expenditure and gift-giving of Herr Jacob Fugger'. Items such as 'golden rings with precious stones, rubies, diamonds, sapphires, turquoises, necklaces, pearls, silk garments, damask, camel hair, and other gems'[18] were showered on dignitaries, investment in the firm's future success.

As the century wore on and *Kunstkammer* became more fashionable, intricate mathematical instruments, unusual works of art and rare natural wonders were added to the list. Among the gifts the Fuggers proffered to their clients and patrons, information was the most useful. With their immense network of offices throughout Europe and beyond, they could listen to the whisperings of every major player on the stage – they were the first to hear if a fleet was captured by pirates, whether a new mining venture would be successful, or an old pope would survive. Jakob Fugger sent regular newsletters to Duke Georg of Saxony and the firm used the established imperial system of messengers, supplemented by their own envoys and merchant convoys. The ultimate fixers, they provided introductions and connected experts with investors, usually on the proviso that they would also have a piece of the pie. In this age of limited specialist knowledge, this was an extremely valuable resource, one the Fuggers were supremely good at exploiting.

The web of the Fuggers' mercantile relationships was overlaid by another, interconnected, network of scholarship. The Fuggers indirectly funded art and scholarship through their loans to the Habsburgs and other noble families, but they also participated directly in patronage themselves. Jakob Fugger commissioned Dürer to paint his portrait in 1518. They had got to know one another in Venice over a decade earlier in the Fondaco dei Tedeschi, the German merchant foundation just by the Rialto Bridge. They met up again at the Diet of Augsburg in 1518. Dürer had been summoned as part of the delegation from Nuremberg, but it wasn't until 1520 that he was able to finish the painting, which now hangs in the city museum. Like Mark Zuckerberg in his trademark sweatpants and t-shirt, Jakob's clothing is deliberately understated. His robe is fur-trimmed but otherwise plain and unremarkable (though no doubt luxurious to the touch if not the eye), he wears no jewellery or embellishment, his clean-shaven face intently contemplates the middle distance, his mouth is thin and pursed in concentration. Only his hat, faintly embroidered with gold thread, hints at his strongrooms brimming with coins, jewels and pieces of silver.

Like entrepreneurs in any age, the Fuggers' success was not only down to personal talent, energy and diligence, although Jakob had all three in abundance; they were also blessed with the right external conditions. In the decades surrounding 1500 there was an upsurge in metal production fuelled by a dramatic increase in demand – copper was needed for weapons and cannon, among other things. This was made possible by technological advancements in extraction and processing, which the Fugger Company invested in heavily, financing new smelting works which were soon producing vast returns. They employed mining experts and local fixers who could smooth the path of business and ensure they achieved the best terms. Lending money to nobles in the Tyrol, Hungary and Slovakia guaranteed them permissions and tax exemptions, and they were often paid back with direct access to the silver and copper, which they were allowed to sell on the open market – until the turn of the century this usually happened in Venice. They constructed roads as part of a huge, pan-European distribution network that carried their cargoes from southern Germany to Venice, either by road via Wiener Neustadt or by sea from Trieste, and north to the Baltic ports of Gdansk and Lübeck, from where they sailed west to Antwerp or east to Russia.

The firm's office in Venice was still doing good business, but they were increasingly taking the copper and silver north-west up the great rivers of Germany to Antwerp. This city on the river Scheldt was becoming the main port for the Portuguese spice trade; ships fragrant with pepper and cinnamon sailed up the Atlantic coast straight past Lisbon and on to the Low Countries, where a sinuous network of rivers and canals carried consignments to every part of Europe. Antwerp's other advantage was its proximity to the sheep of England – the city was the main distribution hub for woollen cloth, a staple commodity which, along with barrels of salted herring, merchants couldn't sell quickly enough. The Fuggers had an office there from at least 1493, and it flourished as Iberian merchants exchanged their spices for German silver and copper, and English cloth. As the Venetian market for metals diminished, Antwerp took over. In 1503 no less than forty-one shipments of copper sailed into the Scheldt harbour from Gdansk; between 1507 and 1526, around half the metal extracted from Hungarian mines was sold there. The Fuggers settled in and bought a house right in the centre of town. Dürer, visiting in 1520 on his ill-fated mission to find a whale, noted that it was 'constructed altogether new and at great expense, with a particular tower, wide and large, and with a beautiful garden'.[19] Spending money in Antwerp, as Dürer discovered, was all too easy; coin slid through the fingers, in exchange for a multitude of temptations: beautiful things, strange things, things that couldn't be had anywhere else – parrots in a cage, pieces of coral, giant's bones, a baboon.

The city quickly became a leading centre of banking too. In 1531 the New Exchange was built to replace the old one in Bullinckstraat; inside its Gothic courtyard you could hear voices from all over the world dealing and haggling, rumours abounded and everyone had an eye on the Scheldt, which cargoes were arriving, who would benefit. Debts were often settled with a combination of cash, precious metals and other valuables – the Fuggers were pawnbrokers par excellence, servicing the needs of the great and the good. In 1523 a weaver from Brussels paid off part of his debt with seven magnificent tapestries, which the firm took to Spain to sell to Charles V. Supreme fixers, they supplied their clients with anything they desired: weapons, furs, opulent textiles and jewels. It is impossible to imagine the sixteenth-century patronage system without them, and their money.

Dürer returned home to Nuremberg, weakened from an illness he had caught on his travels and disappointed he had neither secured Charles V's patronage nor encountered a whale. The city was still a great centre of production, with inspectors scrutinising the output of city workshops. Pewter items for export could only consist of one part in ten lead, compass boxes had to be made of boxwood or ebony, combs from bull horn only. Anything that did not come up to scratch was destroyed, no compensation paid to the humiliated craftsmen. This was a potent incentive for excellence, resulting in the city's unparalleled reputation for manufacturing, which in turn created great prosperity.

As Holy Roman emperor, Charles V ruled over all of these cities – Antwerp, Augsburg, Nuremberg and many more across his sprawling empire – but there was a new threat to his authority, and it was growing. Two years before his accession to the imperial throne in 1519, the Reformation had begun in Wittenberg. As Luther's revolutionary ideas spread quickly through Habsburg territories in Germany and the Low Countries, Nuremberg was one of the first places they took hold, in spite of the fact that Charles had re-established the imperial Governing Council and Chamber Court in the city. He had also followed the tradition of holding his first meeting of the Estates as emperor there. Embarrassment and unease followed when the edict proclaimed against Luther and his followers (Dürer among them) at the Diet of Worms was posted for enactment – many leading figures of the city suddenly found themselves on the wrong side of the law and the city council ignored its stipulations to ban Lutheran texts and persecute adherents. In fact, Nuremberg had enjoyed relative independence from the papacy for some time. The Council had managed many ecclesiastical matters since at least 1500, making it all the more receptive to Luther's reforms.

The struggle for power between the two faiths had begun, and over the following decades the Habsburg Empire would begin to crumble as fifty imperial cities turned to the new religion. Skilled craftsmen were still making the journey from all over Europe to settle in the fabled centre of technology, and some now came to escape religious persecution as well. In 1534 a goldsmith from Vienna called Wenzel Jamnitzer became a citizen of the city. Like Dürer, who had died four years previously, he combined art with technology throughout his career, but rather than painting plants and animals, he cast them in metal, creating 'flowers and grasses that are so delicate and thin that they move when one blows on them'.[20] He was also an expert instrument maker and published a book on the theory of perspective in 1568, with engravings of the five Platonic solids by Jost Amman, illustrator of *The Book of Trades*.

In 1562, Jamnitzer commissioned a portrait of himself. However, unlike Amman's goldsmith in *The Book of Trades*, he is not painted holding the tools of his trade. In his left hand is a silver measuring scale he made himself; in his right, a variable proportional compass – both mathematical instruments rather than metal working ones. Similarly, his tomb features images of arithmetic, architecture, perspective and geometry, and an early seventeenth-century engraving shows him seated at a table opposite the mathematician Johann Neudörffer, discussing perspective. The Nuremberg skyline is their backdrop. Jamnitzer uses a perspective machine, Neudörffer holds a pair of compasses, and there is a globe beside them – in presenting himself as something more than just a craftsman, Wenzel was taking the mantle directly from Dürer, continuing his crusade to elevate the status of artists, scholars and artisans. His emphasis on the scientific aspects of his career shows how it was developing during the century, and with it, those who practised it.

Not only was Wenzel Charles V's favourite craftsman, but his work was loved by the three succeeding emperors too. He stayed in Nuremberg until he died in 1585, resisting imperial commands to move to the court in Vienna or Prague, but continuing to make beautiful, ingenious things for his patrons. He made works of art and instruments for Charles, who was interested in maths and astronomy, and enjoyed looking up at the night sky when the demands of state allowed. Thanks to Regiomontanus, Dürer and Jamnitzer, Nuremberg was the first place in northern Europe where the combination of commercial success and technological ambition came together to create a new world of knowledge, an inspiring example to others; the city remained a thriving centre of instrument making, but this example too was beginning to spread to other places. One of these was a small town to the south of Antwerp, in the Habsburg Low Countries, another place Charles V looked to for specialist instruments.

2

LOUVAIN

THE WONDERS OF THE WORKSHOP

In May 1547, John Dee left his student digs at Cambridge and took a boat across the Channel to the Low Countries 'to speake and conferr with some learned men, and chiefly mathematicians', at the University of Louvain, home to two of the most original scholars and instrument makers of the day: Gemma Frisius and Gerard Mercator.[1] He later complained that the lack of mathematical and astronomical teaching at Cambridge had forced him to travel abroad in search of enlightenment. At this point, England was not yet much of a feature on the map of knowledge, but change was in the air.

John Christopherson, one of Dee's tutors, organised the trip and even managed to secure funding from Trinity College to pay for it. This was to be the first of many trips abroad during Dee's long life and, whether he came ashore at Flushing, Antwerp or another port, it must have been a thrilling moment for the twenty-year-old student. The Low Countries were England's main trading partner and, with a good wind behind you, the crossing only took a few hours. It's hard to believe Dee would not have passed through Antwerp on his way to Louvain, which lies a few hours' walk through the gently undulating countryside to the south-east. Echoing Dürer, he wrote later that the city was 'the emporium of all Europe', a place where you could buy anything, from diamonds and guns to cinnamon and secrets. Antwerp was regularly compared to Venice and, like its counterpart on the lagoon, this northern melting pot became a magnet for printers as well as merchants. One of the greatest booksellers of the age, Christophe Plantin, newly arrived from Paris to set up business, wrote to Pope Gregory XIII that 'in my judgement, no other city in the world could offer more facilities for practising the trade than this one.'[2] He remained there for the rest of his life, running one of the most successful publishing houses of the age, under the sign of the Golden Compasses and the motto 'By labour and constancy'.

At this point, the Low Countries comprised seventeen provinces ruled by the Spanish Habsburg dynasty under Emperor Charles V, whose court was in Brussels. Officially, the population were Catholic, like their Spanish overlords, but the belief in new Protestant theology was growing and starting to have a profound effect on all aspects of life in the region. The Dutch were not thrilled to be under Habsburg rule, but it did strengthen their connection to newly discovered parts of the world and the riches brought by merchants to Antwerp docks, far better connected to the network of European rivers than any port in Spain. They came to exchange their spices for metals from the mines of southern Germany, cloth from England and furs from the east; this explosion in commerce fuelled cultural and social development on a large scale, creating the wealthiest, most dynamic area in all Europe. Italian travellers commented that if Antwerp was the northern Venice, Louvain, with its university and tradition of scholarship, was the northern Padua. In the early decades of the sixteenth century, the city was home to pioneers in developing disciplines like instrument making, cartography and meteorology, paid for and inspired by the discovery of the New World.

A map of Louvain made in 1550 shows a circular city wall punctuated by gates – the western road leading to Brussels, the southeastern one to Liège, and north-west to Antwerp. The streets are intertwined with channels of the river Dyle, and the spaces behind the rows of neat brick houses are a miniature patchwork of fields and orchards. The narrow streets at the centre converge on the Grote Markt, overlooked by Sint-Pieterskerk on one side, and the town hall on the other. Tall, narrow and extremely ornate, this building, which was constructed in the mid-fifteenth century, resembles a wedding cake made of lace. Hundreds of craftsmen must have been involved in creating it and, unsurprisingly, funds ran short towards the end of the project, so the plan to have statues in each of the 236 niches in the walls had to be shelved. They remained empty until the mid-nineteenth century, when a plan was launched to commission statues of important locals, saints and rulers: Gerard Mercator and Gemma Frisius among them. Today they are all filled, with one exception: King Leopold II's statue was removed recently because of his role in the atrocities visited upon the people of the Belgian Congo in the eighteenth century.

Together with the church of St Peter, this was a dramatic statement of Louvain's status as a centre of commerce and culture, towering over the great marketplace that underpinned its financial success. Houses in the city centre are made of brick, as they were in the sixteenth century, although most of them had to be reconstructed after the devastation of the two world wars. The bricks, laid in patterns, are the same dark, dusky pink as the surrounding earth, and the buildings are usually two storeys high with a third attic floor in the characteristic stepped gables. Paintings of the period show shops on the ground floor opening onto the street, with small, tiled awnings to protect from the weather and merchandise laid out on tables underneath; this is particularly evident in paintings by Pieter Bruegel the Elder (1525/30–1569), who painted everyday life in his homeland with such joy and vibrancy, often as a setting for religious devotion, moral questioning or humour. He opens windows onto the sixteenth century, in all its strangeness and familiarity. In his images, people dance and kiss and drink and ice-skate – everything those around Pieter would have done. According to his first biographer, Bruegel and his friend Hans Franckert (who incidentally came from Nuremberg) 'often went out to the country folk whenever there was a fair or a wedding' to get inspiration for paintings.[3] The peasant women wear spotless aprons, snow-white headscarves, there are splashes of jolly red and shiny blue, and they fight and wail and grin. Elsewhere his imagination runs wild with nightmarish monsters and horrific suffering: Lent fights Carnival, the Deadly Sins are ghoulishly brought to life, fiends writhe, revealing the terrifying underbelly of early modern Christian life. The details are irresistible: children munch waffles, the sun glints on baskets of cherries, a lurcher sneaks bread off a plate; this is as close to the past as you can get.

Today Louvain (Leuven) University is spread across a wide area with some campuses a bus ride from the centre, but in the sixteenth century it was concentrated in the centre of town, encircled by the ancient city walls. Students and scholars lived cheek by jowl in this tangle of narrow, cobbled streets, meeting in the communal dining halls and chatting over beer brewed at Brouwerij de Hoorn (founded in 1366, now home to Stella Artois, the biggest brewery in the world). Everyone knew each other and, if you had a particular interest in something, it was easy to find a group of like-minded souls to study and collaborate with. The city's wealth was originally founded on producing linen (in some medieval texts the material is referred to as 'lewyn'), but in the fifteenth century the old medieval Cloth Hall was taken over by the university and used as a classroom. The town was changing.

When Dee arrived, Louvain University had been educating young men for a little over a century. Known as the 'Athens of Belgium', it had grown quickly and was now second only to Paris in reputation. Having completed the traditional BA degree, the three main MA subjects on offer were theology, philosophy and medicine. Students came from all over Europe; one third were from outside the Low Countries. (This tradition has continued; today the university attracts people from 170 countries.) Louvain was a major centre of humanism in northern Europe, not least because the scholar Desiderius Erasmus had taught there and helped found a Trilingual College for the teaching of Latin, Greek and Hebrew using a bequest by his friend and fellow humanist, Hieronymus Busleyden. The enlightened establishment provided lectures free of charge and offered scholarships to students and a professor of each language. This philanthropy was not universally welcomed, however – the university arts faculty, which also taught Latin, did not appreciate competition, and made life difficult for the new college. This did not stop the hundreds of students flocking to soak up the humanist atmosphere – even though the auditorium was extended in the 1520s, some lectures still had to be repeated each day to accommodate everyone who wished to attend. Opening education in this way transformed Louvain into a centre of intellectual brilliance, a place where innovation could flourish.

In 1523 the university's status was further enhanced when Pope Adrianus VI, who had been a professor there himself, established the Pope's College for the teaching of theology; in the following decades eight more were founded. The ideals of communicating clearly in concise, elegant Latin proliferated and producing accurate new translations of ancient texts spread beyond the traditional humanist disciplines into the more scientific arena of natural philosophy via the scholars whose education had been imbued with Erasmian ideals. This was especially marked in medicine; students at the Trilingual College began applying humanist standards of translation to medical texts. Public attacks followed, beginning in 1529, on the medical professors' interpretation of texts, which had been corrupted during their long journey through the Middle Ages and several languages, namely Syriac, Pahlavi (Middle Persian) and Arabic. Armed with his excellent knowledge of Greek, a scholar called Hieronymus Thriverius went back to the original works of Hippocrates and Galen, and his direct translations revealed errors and inconsistencies that clearly needed to be addressed. Unsurprisingly, the medical faculty did not take kindly to these criticisms; for a while Thriverius taught his anatomy course privately. It became so popular, however, that the university was forced to take notice and appoint him in place of two existing professors.

The year before the attacks began, Andreas Vesalius, who went on to revolutionise the study of anatomy, enrolled at Louvain University to study the arts as an undergraduate. Before long he became interested in the family business – his grandfather had been royal physician to Emperor Maximilian, his father royal apothecary – and, influenced by the atmosphere of curiosity instigated by Thriverius, he began looking for corpses to anatomise. Vesalius and Gemma were friends, no doubt brought together by their intellectual curiosity and a desire to push the boundaries of knowledge. One night in the autumn of 1536 they found a human skeleton outside the walls of Louvain, and over the following few evenings, smuggled it into the city so they could reconstruct it and study it. With this level of dedication, it's little wonder that the University of Padua offered Vesalius the chair of surgery and anatomy as soon as he graduated the following year. In his groundbreaking book on human anatomy published in 1543, he described his friend Gemma as 'famous as a physician, and as a mathematician comparable to but a few'.[4]

Undergraduates had the choice of four residential houses at the university: the Pig, the Falcon, the Lily or the Castle. Accommodation varied depending on your budget, though all were conveniently situated in the centre of town. Wealthy boys like Vesalius had their own rooms, servants to look after them and specially prepared food, while poor students like Gerard Mercator, boarding at the castle, and Frisius down the road at the Lily, enjoyed fewer comforts in shared accommodation and ate at the lowest table in the dining hall. All students were required to follow a rigorous timetable that began at dawn with prayers, followed by classes throughout the day, with short breaks for meals. Expectations were extremely high, and the amount of material covered was intimidating. Students had to work long hours and learn everything by rote, memorising vast tracts off by heart. According to university regulations, the curriculum had to 'sustain the doctrine of Aristotle', the scholastic traditions of learning that stretched far back into the Middle Ages. Arithmetic and music featured only briefly, and any further study of maths or astronomy had to be done privately, outside the confines of official university teaching, usually in the masters' own lodgings. Scholars gathered in 'familia' around specific teachers and Dee, like Mercator had before him, found his way into the group centred on Gemma Frisius.

Gemma and Mercator had wretched childhoods in common. Both were thrown the lifeline of education at an early age, transforming their prospects and enabling them to forge careers in new areas of knowledge. In terms of social mobility, this was an impressive feature of life in the Low Countries during this period. Frisius was born in Dokkum in Friesland, a desolate, windblown place on the northern coast of the Dutch provinces, a world away from the prosperity of Antwerp and Louvain. Farmers herded their cattle across the marshes and fishermen took their chances on the stormy grey waters of the North Sea. Neither of these livelihoods were an option for Frisius, who lost both his parents as a child and suffered from 'crooked feet', making it difficult for him to walk. When he was just six, his stepmother took him to the local shrine of St Boniface on a feast day, apparently causing a miraculous improvement in the strength of his legs. He learned to walk but was fragile and suffered poor health for the rest of his life, dying of 'stones' at the age of forty-seven.

Mercator, four years Gemma's junior, lost his father aged fourteen. The family were very poor, but Mercator was adopted by his uncle Gisbert, himself a graduate of Louvain University, who funded his education and gave him a way out of poverty. In previous generations, Gemma and Mercator might have been sent to monasteries and spent the rest of their lives as monks or priests like so many medieval scholars before them, but by the early 1520s times had changed. A secular education was now an option for bright young men, something that would have far-reaching consequences for the intellectual world. Gemma was sent to school in nearby Groningen where he was able to shine; so much so that, at the age of seventeen, the town sponsored him to continue his studies at Louvain on the undergraduate medical course. Frisius is known today for his work in mathematics, but his primary roles were as professor of medicine and a practising doctor. These were well-established and well-paid occupations, vital for a man whose 'urgent condition of our affairs requires a profitable rather than a pleasant art'.[5] Like all physicians, Frisius used astrology in his everyday medical practice, casting horoscopes and assessing the movements of the heavens to diagnose patients and determine their treatment. These were all skills he learned in the university lecture halls; his other activities took place in domestic settings.

In 1529, aged twenty-one and just one year after graduating his BA, he published a new edition of Peter Apian's astronomical manual of 1524, the *Cosmographia*, 'carefully corrected and with all errors set to right, by Gemma Frisius'. Gemma Frisius had arrived, and from this moment on, the eyes of Europe looked to the Low Countries for progress in geography, cartography and astronomy. Apian's text is a layman's introduction to astronomy, geography and mathematical instruments, which Frisius adapted to make it more even more accessible. In a canny commercial move, he also began making instruments to sell alongside the text. There were very few workshops producing items like astrolabes and astronomer's rings, while books like *Cosmographia* were introducing them to a wider audience, creating a new market. His next move was to design 'a geographical globe with the most important stars of the celestial sphere' – a combined terrestrial and celestial globe. He worked in collaboration with his friend Gaspar van der Heyden, a local goldsmith who did the engraving work. Gaspar was twelve years older and already established as a successful goldsmith in the town with his own workshop; as the son of a local surgeon, he had all the contacts and standing Frisius lacked. He had already made a globe in 1527 with a monk from Mechelen, Franciscus Monachus. The 'gores' (the petal-shaped segments on which the maps were printed before being pasted onto the globes) would have been printed at the publishers in

Antwerp, but pasted and finished in the workshop where the spheres were made and inscribed, 'Gaspar van der Heyden, from whom this work, which cost much money and no less labour, may be acquired.'[6]

Gemma published *On the Principles of Astronomy and Cosmography, with Instruction for the Use of Globes, and Information on the World and on Islands and Other Places Recently Discovered* (like his first book, printed in Antwerp) to go with the globe. In chapter xvii, Gemma described his solution to the problem of determining the longitude of a particular location using a small, portable clock. Three years later, in the second edition, he broadened this to finding longitude at sea, something that was crucial to the progress of navigation. The provisos were that the clock must always be accurate and must not stop, nor be affected by the motion of the vessel it was carried on. In 1529, longitude had been the main topic under discussion at a meeting of leading cartographers and navigators in Badajoz, Spain; this was probably what directed Gemma's attention towards it. It took another 250 years before John Harrison designed such a clock, and Gemma's solution to the longitude problem was finally realised, transforming navigation and cartography.

Although globes had been known and occasionally fabricated since ancient times, in 1530 they were still unusual. People were not used to being able to look at a replica of the planet or the heavens, nor did they know how to use this type of information. It is difficult to imagine, from a twenty-first-century standpoint, how it would have felt to see the whole sphere of the earth for the first time, to hold it in your hands, to examine the continents mapped onto it and spin it around to reveal the quasi-mythical far side of the world. It is perhaps evidence of the success of the Louvain workshops that globes were better known by 1559, when Bruegel painted four of them in his dizzying work *Flemish Proverbs* – one is shown upside down on the wall of an inn, alluding to the topsy-turvy world of the sixteenth century, while someone defecates on it out of the window above. By this time, globes and other specialist instruments were being manufactured across Germany, Italy and the Low Countries, and even in England.

In the early sixteenth century, only a small number of workshops produced these marvellous objects, usually engraved spheres of wood or metal made on commission for wealthy clients. The printing press made a new kind of globe possible, one that was made of two hollow hemispheres, usually of wood but sometimes papier mâché and plaster, glued together with the maps printed on gores and then pasted onto the surface. This type of globe was cheaper and easier to produce, enabling workshops to make them in larger numbers for general sale rather than only on commission, reducing the price and increasing their availability. Gemma saw the potential of this and ran with it. His combined globe, which was being produced in Louvain workshops by 1530, was the first of several he designed, each one with improved geographical information which was constantly being updated by sailors and merchants returning to Antwerp from voyages. In later life, Gemma had his own workshop, probably on the ground floor of the house he lived in with his wife and family, a multipurpose space for study and teaching, writing, designing and making. As orders for globes and the other instruments increased, the amount of labour involved meant they had to employ other local workshops to help with manufacture. Gemma was on his way. He still managed to find time to study for his master's in medicine, although he completed it two years late.

In 1533 Gemma published a new edition of the *Cosmographia*, which went on to be a significant financial success. He added a chapter explaining his new technique of using triangulation to accurately survey an area – a method that is still commonly used today. He also designed an instrument called a plani-metrum to make the process easier. Looking out over the softly undulating countryside surrounding Louvain, Frisius realised that he could work out a location by measuring the angles between it and two points (the baseline) using the distance between them, enabling a map to be drawn to scale. The flat Low Countries landscape was conducive to mapping; all Gemma had to do was climb one of the towers of St Peter's church – which were being built at the time – to see views of Antwerp, Brussels, Mechelen and beyond. Triangulation made it possible, for the first time, to correctly locate places on a map, to capture the vast tracts of the planet and plot them onto the page to scale. The whimsical maps of the Middle Ages like the Mappa Mundi in Hereford cathedral, which, in the absence of geographical knowledge, presented a vision of the cosmos based on imagina-tion and faith, were gradually replaced by accurate charts and surveys.

Maps enabled geography (the description of the world based on observation and measurement) to gradually eclipse cosmogra-phy (the conception of the universe based on philosophy and conjecture), changing the way humanity saw the world and how they approached it as an area of study. The ramifications of this were profound, especially for a society that was voyaging around the planet. As sea routes east to India and west to the Americas opened up, the need to be able to visualise, plot and ultimately, claim, the contours of the expanding world became a matter of supreme urgency, making Gemma's method of triangulation one of the most transformative developments of the age. As the decades passed and exploration became exploitation, resources of all kinds were focused on the race to colonise the newly discovered lands.

Gemma married a young woman from Louvain called Barbara in 1534, and set up a marital home where he 'started giving lessons in mathematics this past Ash Wednesday, and the number of my students keeps increasing every time'.[7] This must have been a sizeable space, large enough to accommodate the pupils, the books, instruments and implements he owned. No evidence remains as to where his house was, however, a portrait made after his death shows him seated at a desk that is covered with tools of various kinds: scissors, knives, compasses, pincers and a rule, among others. He clasps a small celestial globe with confidence and familiarity – they are dexterous hands, used to handling metal and wood, to adjusting, correcting and holding still. Behind him on the wall hangs an intricate astrolabe and other examples of instruments he has designed, created and used, as beautiful as they are practical.

There is little written evidence for the workshops in Louvain in this period, but we can get an idea of how they would have looked from contemporary images. These depict workshops on the ground floor of dwellings, with wooden shutters that could be opened during the day. Oiled linen cloth was also used to cover windows, providing protection from the worst of the weather and letting a little light in. (Glass panes were a luxury only the very wealthy could afford in the 1540s; by the end of the century they were cheaper, and more common.) In the winter, this left people with the stark choice between light and warmth – you couldn't have both with wooden shutters. Candles were expensive and only shed small pools of light, the darkness was always lapping at the edges, and precious few pastimes can be done well in the gloom, certainly not instrument making. Consequently, much more of life happened outside, and people were hardier, better at living with the elements. Inside activities occurred in much smaller, multipurpose spaces. The workshops of Louvain were part of scholars' homes, used for teaching as well as design, study, manufacture and sale, showrooms where customers came to choose and purchase items, to discuss scientific ideas and even to learn how to use instruments. The whole production line was telescoped down into one space, with every part of it on show to visitors.

Sixteenth-century scholars and craftsmen were self-employed, responsible for running their workshops as small businesses with instruments and knowledge as the product, always on the look-out for new patrons and customers. In Louvain, as in most other centres of learning at this time, natural philosophy was studied and the boundaries of knowledge expanded *outside* the traditional seats of learning, albeit by people who had been educated in and were still closely connected to them. This is one of the remarkable features of science before the mid-seventeenth century – it was often carried out adjacent to formal settings by people who were part of those settings but keen to strike out independently with their own research. Gemma completed his MA degree in 1536 and began working as a professor of medicine and a doctor; the mathematical work took place on the side. This was even more true of the emerging disciplines like map and instrument making. The intersection of intellectual knowledge and craft activity was a defining aspect of this period, underpinning many of the developments and inventions in science. Gemma's workshop was a vibrant intersection of these two worlds, a place where new knowledge was propagated, instruments were made, improved and used, and students were instructed. Of the many young men who came to learn from

Gemma Frisius, one would make his mark on history: Gerard Mercator, who arrived at the workshop in 1534.

*

* *

*

The main biography of Mercator, written by his devoted friend Walter Ghim, describes him 'teaching their elements to several private students and from time to time…fashioning and constructing scientific instruments, (i.e. Spheres and astrolabes), astronomer's rings, and similar apparatus in bronze'.[8] For his own education in map-making, Mercator had 'relied only on private direction from the then famous Dr. Gemma Frisius';[9] there was almost no one else with the expertise in the Low Countries at this stage – their intellectual exchange was crucial to scientific development. Frisius made setting up an instrument-making business seem effortless but in truth he was a pioneer in a new growth industry (albeit a small scale one), based on an increasing interest in astronomy and geography fuelled by the exploration and discoveries of the preceding decades. Mercator, who had begun his career pursuing philosophy before moving unsuccessfully on to maths and astronomy (which he studied with Frisius), could see that this new profession would be a more effective way of making a living and providing for a family. He explained:

Since my youth geography has been for me the primary object of study. When I was engaged in it, having applied the considerations of the natural and geometric sciences, I liked, little by little, not only the description of the earth, but also the structure of the whole machinery of the world, whose numerous elements are not known by anyone to date.[10]

The serendipity of a man with Mercator's skills and interests happening to be in the same town at the same time as Frisius and Gaspar, two of the very few people then making globes and maps, is striking. Frisius, Gaspar and Mercator clearly got on well and liked each other – the respect and affection that characterised the relationships amongst their circle is clear. The legend on one of the globes they produced reads: 'Gemma Frisius, physician and mathematician, wrote this work in accordance with certain geographical observations and gave it this form; Gerardus Mercator of Rupelmonde engraved it with Gaspar van der Heyden.'[11] This was an overtly collaborative project, a model for how scientific enterprise could flourish that influenced future generations.

As this legend makes clear, Gaspar was teaching Mercator to be an engraver. This was a highly skilled job demanding a combination of creativity and accuracy, physical talent and intellectual ability. One of the greatest challenges facing every cartographer was fitting so much information onto the surface and making sure it was not only legible but beautiful. Being an expert calligrapher was essential. Mercator had already made several maps by this point and had begun to use an Italian cursive script called *cancelleresca* to mark up place names. Its flowing, slanted shape meant that more information could be fitted in. At the same time, it looked elegant and clean: Erasmus had promoted it for these qualities. In March 1541 Mercator published a book explaining how to write the script, including detailed practical instructions about quills, inks and sharpening tools; it sold well and generated some much-needed cash during the long, labour-intensive globe projects. Soon he too was teaching maths, a measure of how much demand there was for extracurricular studies at the university in this period. Making instruments was another source of income – doubtless some of these were purchased by his wealthier pupils.

Earning money was a constant preoccupation. With no private fortunes of their own to fall back on and growing families to provide for, they had to secure the support of people who did. Fortunately for Gemma and his fellows, the imperial Habsburg court was just a few hours' walk to the east, at the Coudenberg Palace in Brussels.

Patronage wasn't just about money. It was an elaborate, mutually beneficial transaction – noblemen offered scholars opportunities for advancement, social prestige and political protection. In return, a coterie of loyal scholars allowed nobles to cultivate connections and project valuable cultural cachet. Johannes Dantiscus, Polish ambassador to the imperial court and international man of culture, was in contact with Gemma well before 7 August 1531, the date of the first surviving letter between them.[12] That same year Dantiscus ordered a globe from Bartholomew de Grave (Graevius), Gemma's printer in Louvain, who must either have been producing his own globes or selling them on behalf of Gaspar and Gemma.

While there were transactions of various kinds at the heart of these relationships, there was also genuine friendship and love. A couple of years later, Dantiscus tried to persuade Gemma to travel to Poland to work with Nicolaus Copernicus, who was putting the finishing touches to his new heliocentric system of the universe. Gemma was quick to voice his support after the publication of *De Revolutionibus* in 1543, and one of very few of his generation. However, he was not keen on the idea of leaving Louvain, his pupils and his workshop. This was doubtless partly down to his fragile health – in the letter of August 1531 he laments that he is sick with inflammation of the liver, signing off with a piteous request for 'doctors and medicine', saying, 'if I live, I shall not be ungrateful.'[13] He also discusses books he has been trying to obtain for Dantiscus; their relationship is clearly well established, and based on mutual assistance and interests. Another courtier with similar interests was Charles V's chancellor Nicolas Perrenot de Granvelle, whose son Antoine became friends with Mercator at university, giving a level of equality and intimacy to their relationship that would not otherwise have existed. They came from very different backgrounds, but in the lecture halls and brew houses of Louvain, they were, intellectually at least, on the same level, two young men fascinated by the world around them.

Imperial Habsburg patronage was the ultimate goal. Hopes of attracting Emperor Charles V's attention and renewing his pension kept Albrecht Dürer hanging around in the Low Countries in 1520 and 1521, and a few years later, Gemma was presented at court on Dantiscus' recommendation. One of Gemma's first imperial commissions was a new terrestrial globe detailed in a charter dated December 1535. This new globe would 'be improved and enriched and more beautiful than their earlier [globe]' and 'for the general use of enthusiasts',[14] a stunning symbol of the emperor's power and dominion. The charter protected Gemma and van der Heyden's ownership of its design; the only problem now was how they were going to achieve its lofty goals. Woodblock printing was far too clumsy, so they decided to engrave onto copper, which allowed Mercator to achieve higher levels of precision and detail. This was the first of a series of globes he produced during his lifetime, some of which were sold directly to clients from the workshop he set up a couple of years later (he, too, married a Louvainite called Barbara and they settled down together to the north of the city, near the church of St Gertrude); others were purchased by bookdealers and taken across Europe. Like Gemma, Mercator also made instruments like astrolabes and astronomer's rings, usually cast from brass and carefully engraved by hand. By the mid-1540s, the workshops of Louvain were famous for producing the most accurate and the most beautiful tools for studying astronomy that money could buy, eclipsing even the masters of Nuremberg. Mercator made a set for Charles V, who used them as he travelled around his dominions, until they were destroyed in a fire in Ingolstadt. He immediately ordered replacements.

Frisius, van der Heyden and Mercator (and later, Dee) were friends and colleagues, with no boundary between their personal and professional relationships – this division did not exist in the sixteenth century. A household was not the tiny nuclear family it is today, but more often than not a sprawling *familia* of blood relatives, domestic servants, animals, assistants and other employees, the centre of both family life and business. The richer or grander you were, the larger this group would be, expanding to include friends' children, hangers-on, visitors, live-in scholars and specialist workers. The workshop where items were manufactured and sold – be it a smithy, printing house, chandlery or shoemaker's – was usually on the ground floor or adjacent to the domestic quarters. Rembrandt's house in Amsterdam, where he lived early in the seventeenth century, had two showrooms on the ground floor, an entrance hall and an anteroom. The walls were covered with paintings for sale, and clients could also stay the night. There was also a salon which the Rembrandt family would have used as a sitting room and bedroom, a kitchen, where the family ate, at the back of the house, and a tiny office where Rembrandt would have done his paperwork. Upstairs were several studios, including one for his pupils, and a room with shelves full of plaster busts, weapons, shells and other rarities – Rembrandt's cabinet of curiosities, many of which appear in his paintings.

The Louvain scholars' houses were on a humbler scale than this, but they, too, would have gathered in their houses, where their studies and workshops were, as well as their kitchens and parlours. It is likely that they stayed with each other at various times, and certain that they shared workplaces. Their set-up, collaborative mode of working, instrument designs and resulting successes were influential for decades, with scholars from a range of other countries looking to them as a model for scientific enterprise that was intellectually satisfying and financially viable.

* * *

Louvain may have been flourishing intellectually and technologically, but when it came to politics, things were in turmoil. Protestantism was on the rise across northern Europe, the fight with Catholic authorities tearing communities apart. The new faith was informing and becoming integral to a burgeoning intellectual scene, and soon Catholic suppression would cause migrations of small yet significant communities, with far-reaching consequences for the places they fled to.

In 1542 Louvain was attacked by an anti-Habsburg pro-Lutheran coalition with the dread 'Black' Maarten van Rossem at their head and saved by the efforts of its students (Frisius, skinny and pale, was among those on the ramparts). By the time Dee arrived in 1547, the Spanish authorities, led by Queen Maria of Hungary, had begun to clamp down on anything that looked remotely reformist. Lutheran ideas had spread quickly from Germany to Switzerland and the Netherlands, igniting the fires of religious reform in many people who were dissatisfied with the corruption of the Roman Catholic Church. Luther's works and the Bible were translated into Dutch and published in unprecedented quantities – in the twenty years between 1520 and 1540 there were eighty Dutch editions of Luther's works, and only nine in English.[15]

Some of the educated, cultured inhabitants of towns like Louvain, whose freedoms had been eroded by their Spanish overlords, found a lot to like in Luther and his followers' ideas about how a society could be organised. The connection between freedom of worship and political independence became increasingly pronounced. Proud, independent and wealthy, the provinces grew discontented with the old feudal organisation of the Church and the repressive measures taken by the Habsburg regime, bent on stamping out any pockets of Lutheran influence. It was difficult enough to maintain power over the Dutch when both sides were worshipping in the same Church, so the widespread adoption of Reformation ideas made for a dangerous situation, one that the Spanish struggled to control.

In 1531, Charles V had appointed his sister, Queen Maria of Hungary, as governor of the Netherlands. She ruled unwillingly until 1555, regularly trying to resign and often complaining about the difficulties of dealing with her subjects. Although personally tolerant towards Lutheranism, she was forced to adopt her brother's hard-line stance and enforce his religious laws. This severity had the very opposite effect to its intention; Protestantism, and in particular, Calvinism, became increasingly popular, a lightning rod for any kind of discontent. As any society knows, and Catholic ones most of all, creating a martyr can be extremely dangerous. Nevertheless, this is what the rulers of the Low Countries did many times over in the 1530s and 1540s. The authorities collected names of suspicious people, first making lists, then arrests. Judging by the forty-three names gathered in 1544, Louvain was an epicentre of heresy. There were only five on the Brussels list and, incredibly, just one for Antwerp. The forty-three included Mercator and several of his friends (but not Gemma), most of them skilled craftspeople or merchants: booksellers, furriers, architects, a tailor, a glazier, a haberdasher and the town flautist, among others.

When they came for Mercator, he was not at home 'behind the Augustins' (St Gertrude's Abbey, an Augustinian foundation); a tip-off had sent him fleeing to his childhood village of Rupelmonde, a small port on the river Schelde where it joined the river Rupel, just a couple of hours on foot from Antwerp, still less by water. These riparian surroundings did not provide shelter for Mercator for long: he was arrested and imprisoned in the vast, dank bulk of the castle in whose shadow he had grown up. His wife Barbara, back in Louvain, distraught and trying to keep everything together, launched herself into a crusade for his release. She went straight to Pierre de Corte, leading member of the local great and good, and former rector of the university, who enlisted other fellow dignitaries and the university (Mercator was still on the roll so officially under its protection) to protest his innocence.

Mercator was both well liked and respected. De Corte's letter describing his good character was soon in Queen Maria's hands, but she was immovable. He was accused of resisting arrest in addition to the initial charge of heresy, and it was clear his position was extremely dangerous. As the summer wore on, others on the list were questioned, convicted and sentenced. Antoinette Van Roesmaels became the focus of the case, admitting to holding bible readings and believing neither in purgatory nor transubstantiation (that the host is actually transformed into the body of Christ). On 15 June, they dug a grave in the town square and buried her alive. Catherine Metsys suffered the same horrific fate – they are remembered today in a huge painting that hangs in Louvain town hall. Antoinette's other co-defendants were either burned at the stake or beheaded; fortunately for Mercator, she told the authorities he had never attended any meetings. He remained in prison for another couple of months before walking out into the September sunshine and blessed familiarity of Rupelmonde to be greeted by his siblings.

This narrow escape had a profound effect on Mercator, but, for now at least, he remained in Louvain and got back to work with a vengeance – seven months away from his workshop had taken its toll on the family finances, commissions had piled up and there was much to do. The next couple of years were marked by continued religious unease and, so long as he stayed in Louvain, Mercator and his family had to keep a low profile, being mindful of what they said to whom. They knew they were being watched. It cannot have been an easy time, but he made 'numerous mathematical instruments' for Charles V and delivered them safely to the court at Brussels. The irony cannot have been lost on him as he worked long into the night to finish commissions for the very man at whose command he had languished in prison for so many months. On 9 October he wrote to de Granvelle:

Honourable Sir, I send herewith the instructions for the use of the instrument, written while I was under great stress. I had intended sending it to you earlier, but…doubted the wisdom of entrusting it to the couriers. Should Your Excellency wish further information, I beg you so to advise me. I have not yet been able to complete the building of the ring, which I had commenced, owing to the unjust persecution of which I have been the victim. I am now working on it and shall send it you soon, as soon as it is finished… Louvain, 9 October 1544.[16]

This was the Louvain John Dee entered in the spring of 1547, enraptured by his first experiences of a foreign country. He and

Mercator quickly became close, talking for hours about how the stars and planets directly affected life on earth, how to make proper observations of the heavens and the weather. It's easy to see how Mercator, coming out of such a tense, mournful period in his life, would have welcomed the distraction of a young scholar like Dee who, if his writing is anything to go by, was full of the exuberance and optimism of youth. Meanwhile for Dee, meeting someone with Mercator's status and talent, a man at the very cutting edge of scientific discovery, was a formative experience – there was no one comparable in England. The educational Protestant reformer Philip Melanchthon's influence was much weaker there, and while in Germany almost every university had a chair of mathematics filled by a professor who was also an astronomer, in England there was nothing comparable for almost a century. Professorships in astronomy and geometry were not established at Oxford until 1619, at Cambridge even later, so it is not surprising that Dee returned to Louvain as soon as he could the following summer and enrolled at the university.

It was not the official lectures on canon law in the university halls that caught Dee's imagination, but the long evenings at Gemma's house or in Mercator's workshop, handling instruments, staring up at the stars and making observations. His fondness for Mercator is clear:

> it was the custom of our mutual friendship and intimacy that, during three whole years, neither of us willingly lacked each other's presence for as much as three whole days; and such was the eagerness of both for learning and philosophizing that, after we had come together, we scarcely left off the investigation of difficult and useful problems for three minutes of an hour.[17]

They discussed the structure of the universe and Copernicus' theory (published just four years earlier) that the sun was at the centre of the universe with the earth orbiting around it, the controversy that would preoccupy astronomers for the rest of the century. On evenings when the weather was clear, they would gather their instruments and go outside to observe and measure heavenly activity, tracking the planets as they span through the dark sky, looking out for comets and new stars.

Astrology had been under attack for several decades; Mercator and Dee were keen to ground it on a more scientific basis and place it within Copernicus' new cosmographical framework. They were especially interested in finding out exactly *how* the planets and stars influenced events on earth. 'That noble debate formerly carried on between us'[18] during the long evenings resulted in Mercator's celestial globe and accompanying astrological disc of 1551, and Dee's book the *Propaedumata Aphoristica*, published in 1558, when he had been back in England for several years. In this work he explained that the stars and planets emit 'rays of celestial virtue' that arrive on earth in various strengths and at differing angles, with perpendicular beams having the strongest effect.

Dee had already begun building a network during his years at Cambridge, but his time in Louvain took him to new heights. Not only did he have close and inspirational relationships with fellow scholars, he also made the most of his proximity to the imperial court in Brussels, claiming that 'diverse noblemen (Spaniardes, Italians, and others) came...to visit me at Lovayne, and to have some proof of me by their owne judgements'. These included the Duke of Mantua, Don Luis de la Cerda, Sir William Pickering and 'some out of Bohemia...some out of Denmarke unto me, as Mathias Hacus, Danus, Regis Daniae Mathematicus [Royal Danish Mathematician]'.[19] As ever with Dee, we have to allow for a certain amount of self-aggrandisement, while appreciating the sheer exhilaration of this period in his life when he was meeting and impressing people from all over Europe, and the wider world of knowledge was opening up to him. This would have a transformative effect on intellectual life in Britain, as Dee and his peers brought energy and ideas home from the Continent, stimulating the growth of domestic culture.

In Louvain, Dee became friends with Sir William Pickering, the charming, extravagant diplomat who was supposedly the best-looking man in England. Judging by his own writings, Pickering thought Dee was quite handsome himself, describing him as 'A Tall, slighte youthe, lookyinge wise beyonde his years, with fair skin, good lookes, and a brighte colour.'[20] The two spent time together in Louvain and Brussels where William was Edward VI's ambassador to the imperial court. Dee taught him rhetoric and logic, as well as maths and how to make accurate celestial observations using Gemma and Mercator's instruments. Pickering did not remain long in public life, but was nevertheless an important early patron for Dee, someone who strengthened his connections with the powerful Dudley family when he returned to England.

Gemma died in 1555, and Mercator had left Louvain for the peace of Protestant Duisburg over the German border three years earlier, but the workshop continued to thrive under Gemma's son Cornelius and his colleague Walter Arsenius. The number of instruments that survive suggest impressive production levels, and makers across the continent were influenced by the design and quality the city stood for, just as astronomers were enabled to make better, more accurate observations than ever before. Mercator continued his innovative work for another four decades, making maps, instruments and globes that changed the way people saw the world, and his work was also continued by his son.

Louvain's collaborative, progressive model would shine down the ages as an example of what was possible with the right conditions, people and attitudes, influencing future generations of astronomers in diverse ways. The expertise in designing ever-more accurate instruments enhanced the quality of observational data, its usefulness and status. This strengthened the role of instruments in the scientific enterprise; today, technology is so integral, it is no longer possible to draw a line between the two. Modern astronomy *is* cutting-edge technology, and the complex telescopes that empower us to see into the darkest corners of the universe have their roots in the workshops of Louvain, and the standards and ideals that were generated there. As we will see in the next chapter, Dee took everything he saw and learned there back to England with him, before founding his own unique version of the workshops he had been so inspired by in a rambling house near the river Thames, just to the west of London.

3

MORTLAKE

THE UNIVERSAL LIBRARY

In the late sixteenth century, Dee's house in Mortlake was the pivotal centre of knowledge in England, the place where new ideas arrived from the Continent and people came to share them. Unprecedented in so far as it was not part of a university, a religious foundation or the court, it prefigured institutions like the Royal Society that began to be founded in the following century. It was where Dee made astronomical observations, noted down weather patterns, carried out alchemical experiments, held meetings, conferred with angels and wrote books. Those seeking his contribution to modern science need look no further. It was his, and England's, stargazer's palace.

John Dee was a unique figure in his own time, but in the breadth of his interests, which are expressed in his writings and the library he amassed, he epitomises the dizzying scope of intellectual knowledge in this period. The most important autobiographical source we have is the *Compendius Rehearsall*, a wordy (brevity was not one of Dee's strengths) curriculum vitae describing his 'studious life, for the space of halfe an hundred yeeres', along with extracts of his diaries and two manuscript copies of his library catalogue made in 1583 before he left for Poland.[1] This is an unusually rich amount of source material for someone in this period, especially the diaries, which give us a rare view into the private life of this intensely confounding man. The personal nature of some of the entries – he described his children's injuries and maladies in detail, he was interested in his wife's menstrual cycle and noted when she got her periods,* and he recorded their involvement in a wife-swapping incident in Bohemia – were distasteful and confusing to many historians, especially those viewing him through the prism of traditional science. Added to this, Dee spent the last decades of his life pursuing knowledge by talking to angels through a medium or 'scryer'. This was a problematic, marginal activity that caused him serious difficulties; it has only become more problematic over time, as science has moved away from religion. It condemned Dee in the eyes of many historians of science and made him vulnerable to all sorts of interpretations – in the early twentieth century, he was taken up by the occultist poet–mountaineer Aleister Crowley, which did nothing to enhance his credentials.

Many of the sources have survived thanks to the generation of scholars who lived in the decades after Dee died and found him so intriguing. His reputation for magic and the possibility that he had accessed some kind of special knowledge drove the likes of Elias Ashmole and Méric Casaubon to seek out his papers, dig up the fields around his house and question elderly locals who had known him. Ashmole, living a century later, advertised his interest in anything pertaining to Dee. He received letters and fragments of information from various people, hitting the jackpot on 20 August 1672, when he 'received by the hands of my Servant

---

*   He also recorded her pregnancies and miscarriages. Serious study of the female body and its workings is a relatively recent phenomenon; today, Dee's interest appears far-sighted rather than strange. (See Angela Saini, *Inferior*. London: 4th Estate, 2017.)

Samuell Story, a parcel of Dr Dee's Manuscripts, all written with his owne hand'. Mrs Wale, the source of these precious documents, explained that years before, she had gone to buy 'some Household stuff' with her former husband, a confectioner by the name of Jones. They had found 'a Chest of Cedar wood, about a yard and a halfe long, whose Lock and Hinges, being of extraordinary neate worke, invited them to buy it'. It had previously belonged to 'Mr John Woodall Chirurgeon [surgeon]', who probably bought it after Dee's death when his possessions were sold. Twenty years passed before the Joneses decided to move the chest and heard a strange rattle. 'Hereupon her Husband thrust a piece of Iron into a small Crevice at the bottome of the Chest, & thereupon appeared a private drawer, wch being drawne out, therein were found divers Bookes in Manuscript, & papers, together with a litle Box, & therein a Chaplet of Olive Beades, & a Cross of the same wood, hanging at the end of them.' Not realising they were of any consequence, they wasted half of the papers 'under pyes & other like uses'.[2] In 1666 the Great Fire of London raged through their street and the chest was burned but the books were taken out and saved. Mr Jones died, and his wife remarried Mr Wale who, realising the value of the papers, passed them on to Elias Ashmole.

In the *Compendius Rehearsall*, Dee looks back to high points of his career, as well as the moments of disappointment. Dee lived at a time when the secular, scholarly world was in its infancy, not yet professionalised. Men in his position were constantly trying to secure patronage from the nobility and the Crown, playing a complicated game that often left them living hand to mouth. His interactions with William Cecil (later Lord Burghley), Elizabeth's chief minister, and others are a window into how scholarship and research occurred, who paid for it and decided which directions it should take, and which external factors came into play. The Elizabethan government asked Dee to carry out a wide range of tasks during his career – casting horoscopes, consulting on the queen's health, reforming the calendar and reporting on the comet of 1577 are just a few examples.[3] In turn, he made many attempts to interest them in his own ideas and persuade them to fund his trips abroad, with varying degrees of success. These interactions reveal how freelance scholars like Dee operated and deployed their resources, in particular the places they created in which to work – their knowledge bases – where they kept books, instruments and the like, where they made their observations and carried out experiments. In the *Compendius Rehearsall*, Dee discusses his house at Mortlake and its contents, while the surviving fragments of his diaries provide an interesting record of who was working there and visiting in the years after 1577. There is, however, no surviving description of what the house was like inside, apart from a brief interview Méric Casaubon recorded with a local woman who had worked there and the glimpses on offer in Dee's diaries. Frustratingly, none of the numerous people who visited Dee there left a record that has survived.

The house in Mortlake served many purposes for a great many people, but, above all, it was a library – the first in this country that was truly universal in subject matter. Dee claimed that, at its height in 1583, his collection contained around 4,000 books. About 2,300 printed books are listed in the catalogue he made that year, along with 300 manuscripts, so either Dee was exaggerating, or the remaining books were not listed. No theological texts are listed in the catalogue, so these must have formed a different collection, one that would certainly have included prayer books, hymnals and other practical items for everyday worship. This is an astonishing number for one relatively impecunious individual to have amassed in the sixteenth century.

While he was famous for his scientific books, he owned far more works on history than any other subject, including the large number of documents, muniments and charters he had managed to rescue from monasteries as they were shut down by the Protestant reformers. The library encompassed a remarkable range of subjects: Judaism, metallurgy, botany, navigation, the Rhine, horses, money, Norfolk, oils, games, cookery, fossils and the Armenian Church – the relevant index in the 1991 bibliographic study of his catalogue runs to six (huge) pages, and only includes the printed books.

Dee's personal interest in certain thinkers and their theories are also clear. He owned a staggering total of 157 books by Paracelsus, several on Kabbala, and Johannes Trithemius' books about codes and number theory. This unparalleled wealth of knowledge made the house in Mortlake the most important unofficial centre of knowledge in Elizabethan England. Large-scale institutional libraries were founded in Oxford and Cambridge in the following decades (the Bodleian in 1604, the Parker Library in 1575), but it was several centuries before scholars could consult a national collection. Dee's library was a major source of inspiration for all those that followed, as well as other types of academic institutions that developed in the following century.

Dee began building his bibliographic network early – he started buying books and making friends with booksellers as soon as he arrived at Cambridge University aged thirteen and he continued to do this throughout his life. The books he acquired in England fall into three basic categories. There were the printed titles he bought and ordered through London booksellers, the manuscripts and monastic books he was able to borrow, beg or buy from those he encountered on his travels around the country, and the books (whether printed or manuscript) he received from friends and contacts living abroad. During his time at Cambridge, Dee met many of the figures, like William Cecil, Sir John Cheke and Sir William Pickering, who would not only go on to help him acquire books, but also provide him with patronage and employment; they were the lynchpins of a circle of Protestant-leaning men who went on to rise to power under Elizabeth I. His old friend Sir William Pickering, who was ambassador in Paris for some years, was a major source of books (not only for Dee: he also supplied William Cecil). Scholarly friends in Europe occasionally sent him texts, just as he was given or loaned others as presents by English friends and fellow collectors. All in all, there were several ways in which books found their way to the library at Mortlake.

While domestic publishing in England was 'unavoidably insular and parochial when compared with that on the Continent',[4] collectors were able to purchase a wide range of foreign titles from bookshops. Dee's library contained books published in seventy-five different places. Some of these would have been ordered specifically from abroad, via booksellers, especially after 1564 when the Frankfurt Fair catalogue began to be published annually, an enormous shopping list from which bibliophiles could make their selection; Dee owned the earliest known copies of these in England. He was also closely connected with the family of booksellers called the Birkmanns who were an important source of imported books for his library. Large booksellers like the Birkmanns had many contacts, and even other branches (in Cologne and Antwerp, as well as London) in Europe, which facilitated a network of acquisition for the relatively isolated English collectors.

In addition to the books, people came to Mortlake to use and admire Dee's mathematical instruments and specially designed laboratories. As he explains in the *Compendius Rehearsall*, he had brought the first instruments back to Cambridge years earlier, on returning from his first visit to Louvain:

> And after some moneths so spent about the Low Countries, I returned home, and brought with me the first astronomer's staff of brass, that was made of Gemma Frisius' devising, the two great globes of Gerardus Mercator's making, and the astronomer's ring of brass, as Gemma Frisius had newly framed it; and they were afterwards by me left to the use of the Fellowes and Schollers of Trinity College.[5]

These instruments were largely unknown in England in the late 1540s; those Dee brought home in his luggage would have been some of the very first of their kind in this country. Globes were not produced domestically until the 1590s, so the only way to get hold of one, or two (from 1551 onwards the publication of Mercator's celestial globe to go with his terrestrial one of 1541 set the fashion for them almost always being sold in pairs), was to import them from abroad. This could be done either by asking someone in the Low Countries to send them over, or by ordering them from a bookseller in London like Jan Desserans, who purchased several maps and a pair of Gemma's globes from the Plantin Press in Antwerp in April 1568. He offered a pair of Mercator globes as well, but 'could not give these for less than 24 florins'.[6] In the middle decades of the century, they were still extremely rare, usually obtained through contacts with people on the Continent and only by the wealthy who could afford them.

In this regard, Dee was a vital conduit of knowledge and technology from Europe to England. These instruments gave Trinity College significant kudos and an unusual, if not unique, experience for its students. For Gemma and Mercator, it was a good way of entering a new market; they could have been sure of orders from erudite English gentlemen keen to be at the cutting edge of science. As we already know, Dee returned to Louvain as soon as he could the following summer. He stayed there for several years, studying and making contacts in the Low Countries before moving on to France. When he did finally return to England in 1551, he was flying high on the success of his trip to Paris, where his lecture on Euclid was (according to the *Compendius Rehearsall*) so popular that people were forced to peer in at the windows, and he had declined a position as 'one of the King's Mathematicall readers'. His trunk was full of books, his head full of ideas and his address book full of contacts; the world was his oyster, but he was broke. Back in Cambridge, he borrowed £4 from Trinity and wrote a treatise on how to measure the universe, an old astronomical chestnut to which he applied the new skills he had learned from Gemma and Mercator. This work (which does not survive) eventually found its way onto the desk of William Cecil, who called Dee in. They discussed astrology, alchemy and other shared interests, beginning a lifelong relationship that promised more than it delivered – Cecil made use of Dee when he needed him, but felt no compunction about abandoning him when he did not.

For the time being, however, Dee was taken on by the power-brokers around Edward VI. He tutored Robert Dudley, son of the Duke of Northumberland, the 'Lord Protector' who ran the country on behalf of the young king Edward VI. This was a fortunate posting; Robert was Elizabeth's closest confidant in the early decades of her reign. In February 1552 he began working for the Earl of Pembroke, an ambitious soldier who was keen to employ Dee's skills in mathematical astrology and make use of the predictions they promised. Life was full of potential, as Dee set himself up as a court consultant and roving expert on maths, astronomy and navigation, teaching young noblemen how to observe the heavens and use instruments, and producing complex astrological readings for the great and the good. He continued searching for manuscripts to add to his collections when he could afford them and borrowed several scientific books from Oxford and Cambridge colleges. Money was always tight.

In July 1553, Edward VI died at Greenwich Palace. He was fifteen. Northumberland and other leading Protestants made a disastrous attempt to prevent Mary acceding to the throne by crowning Lady Jane Grey instead. Nine days later, she was deposed and Henry's eldest, Catholic, daughter claimed the throne. Northumberland, Grey and several others were executed, and their families and dependents (Dee included) lost their position in society overnight; under the new Catholic regime, they were persons of suspicion who had to keep their heads down and their hands clean to have any chance of avoiding imprisonment and death.

Initially, Dee managed this quite well, especially considering that his father Rowland was arrested early in Mary's reign and only released once all of his property had been confiscated by the crown. Canny Protestants whose will to survive was stronger than their religious convictions swiftly returned to the Catholic Church; this group included another of Dee's patrons, the Earl of Pembroke, who urged Dee to follow suit. Another powerful incentive was the two church livings Dee had been awarded by Edward VI, which could now only be paid to genuine members of the Catholic clergy, so, assisted by his 'singularis amicus' Edmund Bonner, the newly reinstated Bishop of London, he took holy orders and was ordained as a priest in a single day.

This seems extreme to modern sensibilities but goes some way to revealing what it was like when the official religion changed from one day to the next. There were many like Dee who were forced to choose pragmatism over dogmatism, to put personal belief aside in order to survive and get on. There were also a number of people whose religious fervour would take them to the scaffold or the pyre, but they were outliers. There were many more who were not particularly bothered about doctrinal details – it was the same God, after all. They were more concerned about the damage caused by extreme factions than whether the wine actually turns into blood. Evidence at a local, parish level shows that it was the loss of traditions that people noticed most. Robert Parkyn, curate of a village outside Doncaster, lamented that 'on the Purification Day of Our Lady (vz Candylmes Day), ther was no candylls sanctifide, born or holden in mens' handes, as before tymes laudablie was accustomyde, butt utterly omittyde,' while on Ash Wednesday, 'no asshes was gyven to any persons'.[7] These ancient rituals marking out the passing of the years were integral to the fabric of everyday life; people missed them when they were gone.

Dee retained his close contacts with leading Protestants like Dudley, who introduced him to Princess Elizabeth. Dee visited her at Woodstock Palace, where she was living under house arrest. This was a tense period for the country at large, but especially for Elizabeth, who had danced on a knife edge of uncertainty her whole life. At this point, Mary was (apparently) pregnant. A healthy baby would have dashed all hopes for a Protestant restoration and her accession to the throne. Desperate for information, Elizabeth asked Dee to cast horoscopes and predict what would happen. He performed divination at Woodstock and later at the house of Sir Thomas Benger, the Princess's auditor. This was an extremely dangerous thing to do; any kind of question or prediction about the queen's life was treasonous, and there were spies and plots everywhere. Dee returned to his lodgings in London, where he continued conjuring, possibly trying to call up spirits using crystals. On 28 May 1555 there came a knock on the door: he was arrested and his rooms were sealed up under the charge 'suspicion of Magic'.[8]

He was taken with four other men, including Sir Thomas Benger and Henry Carey (key members of Elizabeth's retinue) to Hampton Court, accused of 'calculating the nativities of the King, the Queen and the Princess Elizabeth'.[9] This charge was later changed to conjuring and witchcraft, and although all four men were eventually released, these accusations were to hound Dee until his dying day. Fifty years later, Dee was still trying to refute accusations that he was 'a Conjurer or Caller or Invocator of devils' in a letter to 'To the King's Most Excellent Majestie', written in London on 5 June 1604. After four months in prison, he was released into the custody of his friend Edward Bonner, who had been appointed to verify Dee's religious position and check for signs of adherence to the new faith. This was a singular stroke of luck that would result in a plan for a national collection of books – a passion the two men shared.* In January of 1556, Dee boldly presented Queen Mary, who had only recently been on the point of having him executed for treason, with 'A Supplication...for the recovery and preservation of ancient Writers and Monuments.'

This was an urgent project – 'a most notable library, learning be wonderfully advanced, the passing excellent works of our fore-fathers from rot and worms preserved, [a]nd also hereafter contin-ually the whole realm may (through your Grace's goodnes) use and injoy the whole incomparable treasure so preserved'

---

*   The luck was not all good, however. Dee's time with Bonner marked him as a Catholic sympathiser and exposed him to attacks by Protestant writers like John Foxe.

– provoked by the dissolution of the monasteries in the 1530s. This had been 'the great dividing line; it meant the immediate dispersal of the contents of all the monasteries with the exception of the cathedral priories'.[10] Over eight hundred institutions had been affected, and thousands of books were dispersed, destroyed and lost. The old college libraries of Oxford and Cambridge, many of them part of religious foundations, had also been on the commissioners' list and, before long, they too had fallen victim to the reforming wave that overcame the country in the mid-1530s and continued into the following decade.

Dee did not limit his plans for collecting to England; the Supplication explicitly mentions searches in the great libraries of the Continent. He was already aware that the full breadth of knowledge contained in manuscripts was not available at home, something that was borne out by his own collecting. While the plan was to copy interesting manuscripts and then return them to their owners, Dee's collection contained many books that had been borrowed and never returned. In some cases, this was down to forgetfulness, but in others it was intentional – preserving manuscripts sometimes meant stealing them and Dee certainly wasn't the only collector guilty of this.

In the absence of a Royal Commission (there appears to be no record of a response from Mary), Dee went about achieving the vision set out in the Supplication on his own. In the 1550s he was mainly purchasing medieval scientific texts, because his funds were limited, and these were texts he needed for his work. Dee also owned many historical records and monuments and was interested in the history of England and Wales (his own Welsh ancestry was a particular hobby and source of pride); this kind of text was more difficult to get hold of than new, printed books, although they could occasionally be found for sale in bookshops.

Dee travelled widely and worked hard to get his hands on them, visiting the old religious houses and using all his charm and powers of persuasion. On a visit to Wigmore Castle in Wales, he 'espied a heap of old papers and parchments...there to lye rotting, spoiled, and tossed, in an decayed old chappell' – in the decades following the dissolution of the monasteries there were plenty of treasures to be found by those who knew where to look.[11]

Mary's short reign came to an end on 17 November 1558, when she died aged forty-two. That evening, her closest advisor Cardinal Reginald Pole, Archbishop of Canterbury, also breathed his last, just a short boat ride up the Thames in Lambeth Palace. It must have felt as though Catholicism itself had been struck a mortal blow. In her place, 'a very vain and clever woman',[12] a Protestant, acceded to the throne. Dee's first job for Elizabeth after her accession required his astrological skills once again, but this time there was no danger of arrest or accusations of treason. Robert Dudley asked him to produce an elective horoscope for her coronation day which emphasised the positive omens of that date, a task that involved getting out his instruments and making some detailed observations of the stars and planets, before consulting the almanacs, star tables and copies of Ptolemy's *Tetrabiblos* – the main astrological manual of the age.

Dee's prognostications for a long and successful reign were, by and large, borne out. Against all the odds, the predictions of Nostradamus and the prayers of Catholics and misogynists across the Continent, Elizabeth reigned for forty-five years, seeing off a host of suitors, smallpox, several assassination attempts, and the Spanish along the way. Dee got his first taste of her legendary stinginess when he received the rectorship of a small parish far away in Lincolnshire as payment; this very much set the tone for their relationship. Had he been forced to spend the rest of his life living in Long Leadenham, things would have been very different. As it was, Dee was lucky that his mother lived in a spacious house in a much more salubrious location: Mortlake, right on the River Thames and close to the road that led from London to the royal palaces at Windsor and Richmond. He began spending more and more time there, gradually setting it up as his intellectual and professional base. This location would be crucial to the role both Dee and his house played in the Elizabethan age.

In 1562 Dee set off for an extended journey around Europe – Lincolnshire's attractions were clearly limited. He travelled all the way to Rome, stopping first in Antwerp and Louvain, where he visited old friends, although Gemma had died in 1555 and Mercator had moved to Germany, and spent time in the book-shops, purchasing several Hebrew and Kabbalistic texts. He moved on to Paris where he spent time with the erratic scholar Guillaume Postel, then on to Zurich where he met Konrad Gesner, the famous physician and encyclopaedist, before jour-neying down through the Alps and arriving in Padua on 20 May 1563. From here it was across the lagoon to Venice in June, pursuing his growing passion for alchemy, before travelling down to the glinting city of Ravenna. Urbino was his last stop, before reaching Rome in July where he saw the recently completed basilica of St Peter's, designed by Michelangelo, by now an old man in the last year of his life. Dee returned to England via Graz and Bratislava where Maximilian was crowned King of Hungary, in spite of the fact that it was under Ottoman rule and had been since Suleyman the Magnificent conquered it two decades previ-ously. However, his hopes of attracting Habsburg patronage were dashed; the book he wrote and dedicated to Maximilian did not provoke a response. The trip had strengthened his ties to the continental world of learning, however, providing him with inspiration, books and new friends – he brought all of these home with him.

Back in Mortlake, Dee moved in with his mother and began filling her house with books. Dee described it as 'my poore cottage' but this was for effect; a woman who worked there during the 1590s later recalled 'four or five rooms filled with books' and if we take into account the fact that, in addition to over 2,500 texts, a wide array of instruments and visiting scholars, he lived there with numerous children, a wife and his mother, to say nothing of several servants and assistants, it must have been a fairly large 'cottage'. He also mentions having built and renovated it to make room for three laboratories full of equipment for alchemical experiments and what he calls 'Pyrotechnia', however, these were probably in separate buildings in the garden, given the explosive nature of the activities that took place in them.

The house's location, right by the river Thames and just up the lane from the newly built parish church of St Mary the Virgin, was close to the route regularly taken by courtiers between the royal residences of Richmond, Windsor and the palace of Whitehall. This meant a constant stream of visitors, including the queen herself, who dropped in on him at home several times on her way in either direction, by a strange twist of fate managing to turn up just after Dee's second wife had been laid to rest in 1576 and shortly after his mother's burial in 1580. Many visitors would have arrived by boat, sculled down river by one of the wherrymen who plied their trade on the busiest thoroughfare in Tudor England. Dee's books would have been transported in much the same way, sent down by barge from one of the bookshops that huddled around old St Paul's churchyard in the city of London, or loaded onto a smaller vessel after arriving in the Thames estuary in a shipment from the Continent.

Mortlake was famous for its collection of books, but it was also well known for its instruments, laboratories and marvels like the 'glass so famous' that produced optical illusions which Elizabeth asked to see when she dropped in on Dee on 10 March 1575. Sir William Pickering had left this to Dee in his will. Another 'glass', made of obsidian, now in the British Museum, is part of the small but notorious collection of items associated with Dee. A circular piece of highly polished black volcanic glass, it was made in Mexico by Aztec craftsmen as a symbol of the god Tezcatlipoca, 'lord of the smoking mirror'. If it did indeed belong to Dee, as the inscription on the case by Horace Walpole claims, he might well have obtained it from a Spanish merchant in Antwerp or at the imperial court in Brussels, recently returned from the Americas. The mirror and the 'shew-stone' (crystal ball, also in the British Museum) were both used by Dee to summon angels; he mentions the latter in his library inventory in the *Compendius Rehearsall*, but not the mirror, which seems strange.

Dee also mentions 'Two globes of Gerardus Mercators best making; on which were my divers reformations, both geographi-call and celestiall: and on the celestiall with my hand were set downe divers comettes, their places, and motions, as of me they had been observed' as having been in his library but we don't know when he obtained them. He may well have brought them back with him in 1551, along with another instrument made especially for him by Mercator with 'a horizon and meridian of copper'. Along with the various sea compasses, the clock made by the 'notable workman Dibbley' which could measure seconds and the 'magnes-stone' (a large magnet), this builds a picture of Mortlake as a research centre where different people brought specialist equipment to be used and kept safe, a place where astronomical and meteorological observations were made and data collected. Unfortunately, Dee noted these in a copy of *ephemerides* that did not make it into the hands of Ashmole et al.; had they survived, his legacy might have been less overshadowed by the occult activities that preoccupied him in the final decades of his life.

Dee's inventory also includes details of 'certain rare and exquisitely made Instruments mathematicall', including a five-foot quadrant and a ten-foot *radius Astronomicus*, which was conveniently mounted in a frame so it could be easily used to make 'heavenly observations or mensurations on earth'. Both were designed by Richard Chancellor with whom Dee made observations of the meridian of the sun's height. Chancellor was a navigator who had been introduced to Dee by their mutual patron, Sir Henry Sidney. He led voyages for the Muscovy Company which failed to find the Northeast Passage, but opened trade with Russia and took Chancellor to Ivan the Terrible's court in Moscow. The house at Mortlake was a major meeting place for this group of aristocrats and adventurers, an unofficial headquarters where expeditions were planned, and introductions made. On 23 January 1583 Dee wrote in his diary that 'Mr Secretary Walsingham came to my house: where by good hap he found Mr Adrian Gilbert, and so talk was begun of the north-west straits discovery.' Mortlake was one of the few places where people with similar interests could meet by chance and strike up relationships, Dee's books and instruments at their fingertips. Richard Chancellor seems to have stored his quadrant there because, when he drowned on a voyage in 1556, Dee hung onto it and continued to use it until he left for the Continent in 1583.

Sailors needed instruments, especially astrolabes, to help them navigate and one of these, now in a museum in Belgium, is engraved with Edward VI and the Duke of Northumberland's coats of arms. It was made in 1552 by Thomas Gemini, a founder of the instrument manufacturing trade in England, who was affected by the same religious persecution that pushed Mercator to flee Louvain and settle in the Protestant backwater of Duisburg. As the Spanish authorities clamped down on Reformist ideas in the Low Countries, many people looked across the narrow, grey waterway towards the relative peace and toleration of England. This was a major cause of the spread of new skills and technologies to Protestant areas, helping them develop and become wealthy.

A young instrument maker from a village outside Liège, Thomas Gemini (alias Lambrit) was among those who left in search of a different life. He settled in London near Austin Friars, the monastery Thomas Cromwell took over and rebuilt after it was dissolved, and joined the Dutch Church there, in the nave of the old friary. Gemini was an engraver, producing woodcuts for his own edition of Vesalius' *Anatomy* in 1545 (of which Dee had a copy) and set himself up at a workshop in Blackfriars making instruments alongside his engraving and printing work. In 1555 he printed Leonard Digges' *Prognostications* followed the next year by his *A Boke Named Tectonicon*, the first everyday book in English about surveying, which explained basic mathematics and how to make and use instruments, with the recommendation that they could also be ordered from Gemini's shop. Dee was close to Leonard and even more so to his son, Thomas, whom he cared for from the age of thirteen after his father died. Dee taught Thomas, who went on to become one of the most talented mathematicians of the age, all that he could, especially how to use instruments and make observations. They published jointly on measuring the new star that appeared in 1572 and were still close in 1593 when Thomas loaned Dee ten pounds 'for a whole year'. (A continual problem for Dee was how to monetise the role he had created for himself – goodwill alone does not feed a growing family, and all too often, he and his fellow scholars were promised sums of money which never materialised.) Thomas Digges was the most illustrious of the many students who came to learn how to make astronomical observations using the latest instruments from Dee; bringing this knowledge to England and disseminating it was one of his most important achievements.

An advertisement of 1582 shows how well-established instrument making had become in London. It names several craftsmen and their addresses: 'for metal instruments, Humphrey Cole (North Door, St Paul's), John Bull (Exchange Gate); for wooden instruments, John Read (Hosier Lane), James Lockerson (Dowgate), John Reynolds (Tower Hill)'. By the eighteenth century this small community had mushroomed into the leading centre of instrument making in the world. Cole was the most prolific of this group – many of his instruments survive (twenty-two in various museums around the country) and, with great foresight, he often signed them, making them easy to identify. Originally from the north, Cole settled in London and trained as an engraver at the Mint before branching out into printing and metal working. He produced all the instruments described by Thomas Digges in *Pantometria*, a treatise on surveying and using instruments, begun by his father, which he published in 1571. This work relied heavily on earlier works by Peter Apian and Gemma Frisius, but the crucial difference was that it was in English, bringing this knowledge into everyday life in this country, equipping thousands of people with life-changing skills.

The same sentiment lay behind the 1571 translation of Euclid's *Elements* into English, to which Dee wrote the preface. Written in his usual wordy, overcomplicated style, the overall aim – to bring maths in all its forms to the people in their own language – is laudable. Dee lists the different areas of mathematical knowledge and describes them, but is unable to resist making himself sound cleverer by rebranding many of them with complicated Greek names, contradicting the overall purpose of the project. His personal adherence to the humanist ideal of translation into local languages is confirmed by the number of books in English, French, Italian and other vernaculars in his collection.

The library at Mortlake was unique in Elizabethan England in terms of its volume and disciplinary breadth, but there were other, similar collections being formed around the same time. The closest in both spirit and function was at Syon, the Earl of Northumberland's house at Isleworth, a major centre of scientific activity with an observatory and a kind of Renaissance laboratory. Known as 'the Wizard Earl',[13] Henry Percy was at the forefront of contemporary scientific collection and scholarship, passionate about natural philosophy and an important patron of figures such as Thomas Harriot, Walter Warner and Robert Hues. Percy was influential in the general development of the sciences in the period; he was not merely a patron in the formal sense, but a fellow scholar who studied alongside his clients, opening his homes and collections to them for the greater good of learning. A committed Catholic, Percy was imprisoned in the Tower by James I but even here he 'commonly spent £50 a year on books and was obviously a good customer of many of the leading booksellers of the London of his day'.[14] Like Dee, he was determined to expand the limits of knowledge by whatever means possible.

John Dee stands alone in the Elizabethan scene, however, in several ways. First, as a bibliophile, he was surrounded by men whose wealth and position enabled them to cultivate collections of books for a variety of reasons, be they personal, political or religious. Dee did not fall into this group, nor was he attached to a university. Instead, he set up his own 'academy' at Mortlake on the banks of the Thames, a lone and often lonely figure hovering on the fringes of the court, the Church and the universities, always striving for success and stability, often sabotaging himself with his pomposity and paranoia. With practically no private means, Dee, like Gemma and Mercator, was forced to earn his own, and his growing family's, keep. However, he couldn't convert his talents into a stable income like his friends in Louvain. He failed to secure effective patronage, and he lost the two church livings he had been awarded by Edward VI and Elizabeth. Along with his aristocratic patrons, Dee also worked for various members of the Elizabethan government. The position of the intellectual – or 'intelligencer' in Elizabethan society is a complicated and often obscured one. Dee did not fulfil a formal role and hence there is little evidence to tell us exactly what work he did, how, for whom and for how much.

The jobs he did carry out were usually commissioned by William Cecil, the most influential man in the kingdom. Cecil did not need to wait until the following century to be told that 'knowledge is power'. His position at the head of an enormous network of agents, spies and academics is confirmation enough of the value he put on information. The scope of his influence was extraordinary. He controlled the book trade, taking responsibility for awarding patents from the early 1550s onwards, and there are several instances of his personal intercession. In 1564 he wrote to the Marquis of Winchester

with a special plea on behalf of Arnold Birkmann and Conrad Molyar. They were about to dispatch into England five fatts and two maunds of books from the Frankfurt Fair via Antwerp, not knowing of the proclamation (of 23 March 1564) against trade with the Low Countries. Cecil asked that they might quietly discharge the books – and the four or five boxes of green ginger for Cecil's own use.[15]

With close ties to every major contemporary figure in the book world, and his vast political network, he could obtain a copy of almost any text, or quantity of any spice, come to that, he wanted. The covert nature of this network makes it difficult to reconstruct the details of who was looking for what and why, but his politically motivated collecting was counterbalanced by an intense personal interest in scholarship, the ideals of humanism, alchemy and the occult. There is, however, a letter Dee wrote to Cecil from Antwerp on 16 February 1563 which allows us a glimpse behind the curtain.

Dee begins with a formulaic, obsequious paragraph praising Cecil's 'approved wisdom, with which the Almighty has endowed you' and his 'exact balance of justice…and the natural zeal as well to good letters…as to the honour and public weal of our Country (which now in you freshly flowers and yields fruit abundantly)'.[16] Cecil, it would seem, holds all the cards, but Dee points out that he has chosen to approach Cecil 'among so many others in places of high honour and governance' – there are other players in the Elizabethan power matrix, and Cecil is fortunate that Dee has opened the exchange with him, instead of Dudley or Walsingham or someone else. Dee makes it clear elsewhere in this letter that he is in Antwerp organising the publication of two of his own books, but it is possible that he was also on government business.

In the next section, Dee discusses three scientific texts, bringing the exchange into clearer focus; he has knowledge and expertise which Cecil will be tantalised by. This is clear from Dee's promises of 'Infinite Wisdom of our Creator…branched into manifold more sorts of wonderful Sciences'. Nor has Dee, in his extensive experience, ever heard of an Englishman who is 'able to set his foot or show his hand in the *Science De Numeris formalibus*'. This is probably an exaggeration on Dee's part, but who is Cecil to argue with him? He is at the mercy of the former's expertise, and Dee is clearly hoping to blind him with visions of almost divine, universal knowledge. Dee constantly reminds Cecil of his duty to his country and points to the advantages inherent in the granting of his request. Then Dee begins to discuss the book he has already bought, Trithemius' *Steganographia*, revealing how the knowledge will pass from it, through Dee, through Cecil to the monarch and government at the centre, 'a book for your honour or a Prince, so fit, so needed and commodious, as in human knowledge that none can be fitter or more worthy'. The value of the book is also hinted at: 'A book for which many a learned man has long sought and daily does seek: whose use is greater than the fame spread about it'. The situation shifts and expands again when Dee mentions a Hungarian nobleman who is patronising him in Antwerp, reminding Cecil that Dee's expertise is of value to others, and that he is not the only potential recipient of the information.

It is only at the very end, once Cecil has been put in a position from which he can hardly refuse, that Dee comes to his request. Dee desperately needs more money and more time, although 'God knows my zeal for honest and true knowledge, for which my own flesh, blood and bones would be the merchandise if the case so required.' Here is the final appeal to Cecil's Christian rectitude. He surely cannot let Dee starve to death while on government business. The tone of the letter and the very fact that Dee's position is clearly not an officially maintained one reveals that this instance of power is secret and shrouded, even contemporarily. Dee is not on normal court business – the power he derives is not openly awarded to him; instead, he must titillate, promise and persuade in order to bring about the transaction. It also shows how valuable scholarly information could be in the practical ministrations of government, how scholars could and had to use the promise of knowledge and information as a bargaining tool for patronage and financial support.

An extraordinary range of skills and knowledge were required of the early modern scholar endeavouring to obtain patronage and work. Everything from astronomy and geography through linguistics, rhetoric and logic to mathematics and philosophy could be put to practical use in the early modern political sphere. As the link between academic, theoretical knowledge and its practical application by those in power, these scholar collectors fulfilled essential roles. In this way, the needs of the Elizabethan elite influenced the directions of the development of new knowledge through the scholars they employed and the tasks they demanded of them. Many scholars still pursued their personal research, external to the needs of their employers and the Elizabethan state; however, the ideal that underpinned all scholarly endeavour was the common cause of 'enlightening the population' and improving 'the state of mankind'. This can be clearly seen in Dee's collection of domestic historical sources and his usage of them in developing the idea of a British empire.

By the 1570s the Elizabethan elite had developed a passion for the intellectual complexities and explosive practices of alchemy, and a laboratory stocked with gleaming glass vessels, esoteric medieval manuscripts and sooty furnaces had become a popular status symbol. Lady Mary Sidney (sister of Philip) employed Dee to teach her the basics of alchemy, while at court Elizabeth's interest in occult philosophy and alchemy was widely known, and many books on these subjects were dedicated to her.

Alchemy was one of Cecil's greatest passions too, although this was underpinned by the less exalted and more pressing need to improve the state of England's finances. Transmuting base metals into gold was an all too tempting solution to this intractable problem and scholars who managed to convince Cecil and Elizabeth they were capable of such a marvel found that the taps of patronage turned easily. In 1565 an alchemist from the Low Countries called Cornelius de Lannoy was installed in a laboratory at Somerset House with an annual pension of £120, a fortune beyond Dee's wildest dreams. In the light of de Lannoy's undertaking to produce £33,000 of gold a year, this was a modest investment. In a cruel twist, Cecil and Elizabeth believed his claims because they fitted with Dee's theories in the *Monas Hieroglyphica*.

The philosopher's stone that could transmute base metal into gold or silver was closely linked to another magical substance, the elixir of life, a panacea for all disease and capable of prolonging youth, a miracle that continues to evade humanity while simultaneously persuading us to spend vast sums in its pursuit to this day. Elizabeth funded distilling houses at Hampton Court throughout her reign and paid her servant Millicent Franckwell to distil 'the Queen's medicine' in her privy chamber, in a bid to escape the ravages of age and ill health. Dee failed to attract this level of patronage, in part because he was more interested in the lofty and often baffling mystical aspects of alchemy, rather than making useful chemical discoveries. He was, however, sent to the Duchy of Lorraine to buy specialist glass vessels (imperfect glass was often blamed for alchemical failure), and returned with 'one great cart lading of purposely made vessels, &c' for himself and the queen. These were extremely costly; Dee reckoned the value of his laboratory 'furniture' was over £200 in the inventory of his property made after he returned to England in 1589. In 1592 he petitioned for a position at St Cross in Winchester, listing as one of its advantages its proximity to 'the glasshouses of Sussex...for better matter and shape of glass works and instrument-making'.[17] Like metal instruments, the production of glass vessels was a growing industry, in part fuelled by the popularity of alchemy.

British scholars were connected to their European peers through shared texts, letters and participation in the various intellectual debates which raged across the continent. These links were strengthened and extended by actual visits. The list of European scholars who made the journey to England during the latter half of the sixteenth century is impressive, in quality if not quantity. Girolamo Cardano, Giordano Bruno and Jean Bodin met Dee, and he was already acquainted with the Dutch scholar Abraham Ortelius, who came to see him and his library at Mortlake on 12 March 1577. It's not difficult to imagine how the two men would have enjoyed looking through books together, catching up on gossip and the latest ideas, Dee darting around to find his most impressive tomes to show his friend.

All the major foreign academics who came to England during the later sixteenth century visited Mortlake – the centre of learning had an important reputation. And while the Tudor monarchy may have ignored Dee's requests to set up a national conservation project for the contents of monastic libraries, they were happy to exploit his efforts. His large collection of 'monimentes' was examined by 'divers of her Majesties heralds' and 'other Clerks of the Records in the Tower', who 'satt whole dayes at my house in Mortlake, in gathering rarities to their liking out of them: some antiquaries likewise had view of them'.[18]

Dee was generous with the information he accumulated; no one was turned away and this lack of discrimination made him vulnerable to tricksters. It must have been a huge strain on his wife, Jane, who had the unenviable task of running a large, febrile household on a shoestring. Dee's income could have covered the cost of running the household, but most of it was spent on books, alchemical equipment and assistants before it reached Jane's purse. People were constantly coming and going, noblemen regularly expecting lavish hospitality which Jane had to conjure out of thin air. On 31 July 1583, Robert Dudley (Earl of Leicester from 1564) and the Polish Count Laski invited themselves to dinner but there was no money to pay for it, as Dee was forced to admit to Leicester. In a generous mood for once, the queen sent '40 angels of gold', and saved the day. We can only imagine Jane's relief when the money arrived and she could buy food that was fitting for aristocratic guests.

Jane was born Fromond, a gentry family a few rungs up the ladder from the Dees. She was brought up in the belief that feminine virtue and good housekeeping were the cornerstones of a successful life, but there were times during her marriage when neither was attainable. Married aged twenty-three, Jane was half her husband's age; initially at least, she must have felt quite in awe of him. As time went by, this awe changed into frustration. We rarely hear her voice, but thanks to the frank nature of Dee's diary entries, we do get a real sense of her exasperation, or as Dee's describes it, 'marvellous rage'.[19] Managing the household was a huge challenge – the couple employed 'at least twenty-seven female servants and twenty-nine male servants during the course of their marriage'[20] – and the diaries are littered with references to staff problems: maids setting fire to their rooms (not once, but twice in the same year), a nurse tempted by 'a wicked spirit' throwing herself down a well, and a manservant arriving home drunk and causing havoc. Dee's mysterious activities permeated the whole house, with no firm boundary between intellectual and domestic life. The assistants he employed often had difficult personalities and questionable backgrounds, like the mournful Roger Cook who worked in the alchemical laboratories handling the stills, and they brought their peculiarities to the kitchen table, where they ate alongside the household servants and family. Melancholia had long been considered an essential characteristic of intellectuals and artists, an idea made fashionable by Neoplatonists and humanists who connected it with divine inspiration and genius.

Jane was struggling to hold everything together in the late 1570s, but things were about to take a dramatic turn for the worse. On 8 March 1582, a man knocked on the door who would turn their lives upside down. The same night, 'The sky seemed to be on fire, and to grow red like blood it seemed to spread in different directions...while overhead, the blood red clouds were carried toward the south.'[21] If this wasn't warning enough, Edward Kelly (or Talbot as he introduced himself) immediately began to worm his way into their lives, snooping in his host's papers and audaciously adding to an entry about himself in Dee's diary. In spite of these red flags, the prospect that Kelly could impart divine knowledge and secrets through scrying was too tempting for Dee, who fell headlong under his spell, convinced that convening with angelic spirits was the only route to discovering the knowledge he craved.

Before long, furniture was being moved and chambers repurposed to make space for the spiritual creatures and their master, who was given a room upstairs close to Dee's study, which faced west on the first floor of the house, adjacent to a small oratory for private prayer. The urgent need for privacy was a problem – it would not do to have visitors wandering in when the crystal ball was on the table, glinting in the afternoon light and Kelly was having one of his intense sessions with the angels. On 19 June 1583, this is exactly what happened: 'Tanfeld came rashly upon us, into my study: we thinking that the study door had been shut... came undiscreetly upon us, to our no little amazing and grief, fearing his rash opinion afterward of such things.'[22] Mortlake's accessibility, its great advantage when Dee needed to establish himself in Elizabethan society, had now become a liability. This was something he later dwelt upon when petitioning the queen for the living of St Cross in Winchester, explaining that at Mortlake it was too easy for people to drop in on him.

The tension between scholarly seclusion and the demands of attracting patronage was a common theme in this period. The need for peace and privacy to work was the lament of many a Renaissance scholar. In his printing house in Venice, Aldus Manutius wrote, 'there are two things in particular which continually interrupt my work. First, the frequent letters of learned men...then there are the visitors who come...and sit around with their mouths open.'[23] Aristocratic scholars like Montaigne could retire to their ivory towers: people like Dee could not. For them, having friends and contacts who would provide both encouragement and inspiration, as well as practical assistance in providing introductions, texts and other more banal things like lodgings, was vital. At an international level, these contacts were essential. Without them, a scholar was in danger of becoming isolated and excluded from the wider academic community.*

It is possible to reconstruct the patterns of personal and professional connection from many different types of evidence.

---

* Copernicus was the obvious exception, but he had been introduced to Neoplatonism and other Renaissance ideas during his ten years studying at the universities of Padua and Bologna.

Presentation texts are one such source. The title page of Dee's copy of Peter Ramus' *Prooemium mathematicum* (1567) for example is inscribed with the words, 'Johannes Dee 1567. febr. 20. Londini virum et Amicorum singulare' and signed by Ramus.[24] *Alba Amicora* (friendship books) are valuable sources of information; Abraham Ortelius's contains no less than 240 contributors, Mercator, Plantin and Dee among them, evidence of 'the size and importance of his circle of friends, business relations and correspondents'.[25]

These contacts came in very useful in 1583 when Dee made the drastic decision to abandon his house, his books and his instruments and set off for Poland. Cecil's support had faltered; Dee was broke and encouraged by the Polish nobleman Count Laski's promises of patronage. Laski, a passionate occultist, had spent significant parts of his visit to England staying with Dee in Mortlake. It's no surprise that Kelly was behind this plan, keen to spread his wings and see what opportunities lay out in eastern Europe, using the angels to convince his master that leaving England was the only option. The astrological signs were worrying; they both believed that the great conjunction of Jupiter and Saturn set for later that year heralded the end of the world, or some kind of seismic change, and they certainly weren't alone. Anyone with an interest in astrology was worried about this unusual celestial event, eyes were trained on the night sky in apprehension and numerous books were published – so many that the Pope issued a bull against divination in 1586.

In the absence of the official structures and organisations which have since come to define every area of the academic world, sites of intellectual enquiry such as Mortlake were fundamental to the exchange and development of knowledge. In the years before 1583, the house was one of a handful of places where knowledge was transmitted into the English intellectual sphere. It was a portal through which ideas entered England via books, instruments and their users, who carried the knowledge they contained out into diverse areas of Elizabethan life. It was a site of abstract communication between two geographical areas, where English scholars could find texts and authors which were not available anywhere else in their country, a point of contact between the academic and the practical spheres, and a place where like-minded scholars of various nationalities could meet each other personally. By the summer of 1583, however, Dee had become convinced that his future lay elsewhere. The library at Mortlake may have been a thriving intellectual centre, but it was not providing him with enough to live on. He had a catalogue made of his books and chose which ones to take with him, then packed up his personal belongings and the show stone. On 21 September, he and Kelly set off with Laski; their wives, servants, children and 800 books trundling on wagons behind them, into the unknown. We will re-join them later in eastern Europe, but first we are off to Germany.

4

KASSEL

BALCONIES AND GARDENS

The Landgraf Wilhelm appears to have transported Kassel to
Alexandria.

Petrus [Peter] Ramus, *Scholarum mathematicarum libri unus
et triginta* (Basilae, 1569)

Kassel nestles in the rolling hills of central Germany, on the road leading up from the Alps to Hannover and Hamburg; Leipzig and Dusseldorf are equidistant to the east and west. It is central yet isolated, lying in the broad Fulda river valley, the great cities of modern Germany forming a distant ring around it. The initial settlement, first mentioned in 913, was a manor. By 1277 there was a castle and a bridge over the green water of the Fulda. The huge expanse of the plain gave settlers a good view of approaching enemies, while the river gave them protection and transport. By the sixteenth century, it was an important stopping point on the trade routes from Mainz and Frankfurt to the north up the Weser river, east west between Cologne and Leipzig, ruled by Philip I the Magnanimous from 1509 to 1567 and then by his son, Wilhelm IV, who reigned from 1567 to 1592 and made the city famous as a centre of astronomy and mathematics.

Today it is a small, unassuming town built on a massive scale. The streets are as wide as motorways, the buildings monumental, as if designed for giants. Traffic lights preside over vast intersections providing scant comfort to pedestrians scurrying to the safety of the distant pavement. There is hardly anything left of the city its sixteenth-century inhabitants would recognise; just the odd church remains as a witness to the past. On the top floor of the Stadtmuseum, there is a scale model of Kassel showing the devastating destruction caused by RAF bombing raids during the Second World War. Every blown-out window, burnt-out building and collapsed roof is rendered in agonising, minute detail. In contrast, the maps and paintings of sixteenth-century Kassel appear idyllic, all charming half-timbered houses and rosy-cheeked locals. This is an illusion, of course. In the mid-1520s the Peasants' War had brought mass slaughter to German lands and religion was shaping up to be the great disrupter of the following decades. An early centre of Protestantism in the Catholic Holy Roman Empire, Kassel's position was perilous. One of Wilhelm IV's first acts as landgrave was to improve the fortifications, visible on surviving maps as sturdy walls and ramparts jutting out into the moat that surrounded Kassel on three sides, the fourth protected by the river. These would have been useless against the airborne bombs of the 1940s, but were good insurance against the land-bound Catholic armies of the Holy Roman Empire and made Kassel one of the strongest fortresses in Europe, and one of the few that would survive the brutal Thirty Years' War in the following century.

At this point, Germany was a patchwork of small states and principalities ruled over by local dynastic families, whose numerous intermarriages created a chaotic web of familial connections. In recent decades these dukes and landgraves had been expanding their horizons beyond the orthodox aristocratic habits of hunting and feasting. Culture and education, initially imported from the Arab world via Italy and Spain, and recently codified into new Protestant forms, were taking hold. The Lutheran scholar Philip Melanchthon (1497–1560), Martin Luther's right-hand man, introduced an educational programme to the University of Wittenberg, birthplace of the Reformation and the engine room of the intellectual revolution that followed, that was emulated across Germany and beyond.

Melanchthon promoted the study of mathematics, astronomy and geography, and was very influential among the German ruling elite, who educated their sons according to his advice. As bony and balding as Luther was fleshy and hirsute, their outward differences belied their shared inner mission to tear down the Catholic Church and rebuild it from scratch, or rather from the New Testament, which they believed should be at hand, in all languages, for everyone to read. As Dürer put it, underneath the engraving he made of Melanchthon in 1526, he 'was able to draw the living Philip's face, but the learned hand could not paint his spirit'. Indulgences, monks, incense, relics and transubstantiation were all on the clearance list, along with purgatory, and the Pope.

Given that no one had a more profound influence on learning in the early modern period, it is astonishing that Melanchthon is so little-known today. Martin Luther's beefy frame dwarfs him in contemporary portraits, just as his legacy has done ever since. The town of Wittenberg is now prefixed 'Lutherstadt', while the institution Melanchthon devoted his life to is called the Martin Luther University of Halle-Wittenberg. Although very different in character, their partnership was the most significant of the century. Luther wrote, 'I am the rough pioneer who must break the road; but Master Philip comes along softly and gently, sows and waters heartily, since God has richly endowed him with gifts.'[1]

Wittenberg University was unusual because it was founded by Frederick the Wise, Elector of Saxony, in 1502, rather than by the Church. Geographically remote, this modest city on the river Elbe north of Leipzig lay outside traditional spheres of influence, giving it a degree of freedom that would inspire later foundations as learning became increasingly secularised – the ideal place to launch a religious revolution. Melanchthon arrived in 1518, aged twenty-one, on Luther's request. On 29 August he gave his inaugural lecture explaining his vision for education to an astonished audience of senior academics. He took up a professorship in Greek, and before long had overall charge of the intellectual direction of the whole university. You can visit his house in Wittenberg, walk through his study and imagine him at his desk, pen in hand, warming himself by the green-tiled stove, theological conundrums unknotting in his nimble mind. In portraits by Cranach and Holbein he looks soulful, kind, with high cheek bones and a sensuous curling mouth – the face of a poet rather than a theologian. Physically, he was described as small and weak, with a classic scholar's body and fragile health; Luther, on the other hand, looks like he might have had a weekend job as a bouncer.

At Wittenberg, Melanchthon designed a curriculum that embedded Protestant theology deep in the minds of its pupils, training them to think in innovative ways, to read the scriptures and focus on praising God through studying his creation. Instead of the dense, wordy commentaries of medieval scholasticism, students approached Aristotelian natural philosophy armed with practical new textbooks (many written by him) that asked direct questions and provided lucid definitions. As they went out into the world, this new generation of teachers, churchmen and royal advisors became the scaffolding on which Lutheran society was built in the second half of the sixteenth century, as Melanchthon's reforms were emulated in schools and universities from Bohemia to Denmark, responsible in no small part for the new religion's success.

As far as we are concerned, Melanchthon's attitude towards mathematical subjects was the most important element of his educational reforms. He established two chairs of mathematics at Wittenberg and viewed the practical sciences as the route to understanding the wonder of God's creation, alongside a more traditionally humanist belief in the value of returning to the original ancient texts equipped with linguistic excellence (like Erasmus' college in Louvain, scholars at Wittenberg were taught Latin, Greek and Hebrew). He was particularly interested in astronomy and astrology, explaining in a letter to a fellow professor of Greek,

> Who is so hard and insensible that he, when looking at the sky and the splendid luminaries, would not admire their changing positions and motions and not wish to know the reason for all this? Human intellect must be lacking in those who are not impressed by their beauty and the attractiveness of knowing more about them. Plato aptly remarks that man was given his eyesight for the sake of astronomy. There is indeed an intimate relationship between the eyes and the stars.[2]

In his *Oration in Praise of Astronomy* (1553) he emphasised the importance of looking up: 'there is no doubt that the sky is so adorned with beautiful lights, and likewise the order of the revolutions is so skilfully planned that through this very beauty and order we are invited to such observation.'[3] Universities across northern Europe followed suit. Thanks to him, thousands of young men were now crowded onto narrow wooden benches, learning mathematics and astronomy, being urged to gaze heavenwards and study the world around them in new ways. This influence extended to the ruling elite too. In 1524 Melanchthon met Philip I, ruler of Hesse-Kassel. Soon afterwards, Philip, who was just twenty, converted to Protestantism, one of the first sovereigns to do so. This was a decision he had been considering since attending the Diet of Worms in 1521 where Martin Luther formally revoked the Catholic Church and refused to recant his *Ninety-Five Theses* in front of Charles V. Philip met Luther there and discussed religious matters with the numerous churchmen and rulers in attendance, among them the future Christian III of Denmark who also converted and brought the new religion to Scandinavia.

Philip set about introducing the new faith to his subjects in Hesse. He was a devout and vociferous supporter of the new religion in every way, with the notable exception of sexual morals – Luther described him as constantly living in 'a state of adultery and fornication', not ideal for someone trying to establish a new type of Christianity. He was unfaithful to his wife Christine of Saxony (1505–1549) within weeks of their marriage yet, despite finding her physically unappealing, managed to have ten children with her. In 1540, he took the extraordinary decision to marry one of his sister's ladies-in-waiting, making him the most notorious bigamist in Christendom. This was a disaster for the Protestant cause, scandalising Hessian society and delighting Catholics everywhere. Philip was furious and did not seem to have either predicted or understood the terrible effect this marriage would have on his lifelong project of establishing Lutheranism. He fathered numerous children with his second wife too, and probably a few others besides.

Wilhelm was the eldest son of this extensive brood. He was educated at court according to Melanchthon's principles, studying maths and astronomy alongside more traditional subjects. Then, in

1546, Philip sent him to Strasbourg to continue his learning away from the dangers of war which were threatening Kassel. He hoped that Wilhelm would be inspired by further instruction in theology, a subject he did not share his father's enthusiasm for. Little information survives about this period, but Wilhelm probably spent time with Philip's close friend, Martin Bucer, whose scheme for religious toleration later underpinned Wilhelm's position and his attempts to bring harmony to the various Protestant factions. In terms of learning, it was mathematics rather than religion that caught the young man's attention in Strasbourg. He attended classes at the city academy with Christian Herlin, who had recently been commissioned to build an astronomical clock for the cathedral. This was Wilhelm's first serious encounter with instrument design and, given his later activities, it must have been formative.

On 14 April 1547, Wilhelm's period of study ended, the start of a pattern that would repeat itself throughout his adult life, as political duty tore him away from intellectual investigation. Back at home in Kassel, the Protestant Schmalkaldic League led by his father were facing defeat by Charles V. They were forced to capitulate, and Philip was taken into custody by the Habsburgs, where he remained for five years. This propelled young Wilhelm out of the schoolroom and straight on to the fraught political stage of central Europe; he had to grow up very quickly, especially after the death of his mother in 1549, when he became sole regent. It was a role he took on with great skill, focusing his energies on maintaining the alliance of Protestant rulers and at the same time pushing for his father's release.

This baptism of fire culminated in Wilhelm leading his troops into battle against the Habsburg Empire at Innsbruck, which they occupied with other members of the alliance and crucial support from the French. Philip's release was negotiated as part of the resulting Treaty of Passau, under which Charles V agreed to Lutheran religious freedoms and relinquished the aim of restoring Catholicism. This was an impressive outcome for the twenty-year-old, who, probably with a sigh of relief, handed back the reins of power to his father and returned to his studies. He did, however, carry on his involvement in government and regularly consulted on matters of state, enhancing the reputation he had gained as an intelligent, moderate figure in German politics, someone who could be relied upon to mediate and make peace – qualities that would stand him in good stead. From now on, his position as a respected leader among his fellow Germanic princes was assured.

*   *
*
*   *

For the next few years, Wilhelm focused on astronomy and mathematics, the subjects he found most compelling. He probably began by studying Sacrobosco's *De sphaera* before moving on to Peuerbach's classic astronomical text, *Theoricae Novae Planetarum*. So far, so traditional, but Wilhelm also made his first observations around this time, in 1551, with Eberhard Baldewein (1525–1593), his father's *Lichtkämmer*, in charge of heating and lighting at the royal palace. This was the beginning of a valuable partnership that resulted in the construction of a great many instruments as Baldewein's roles expanded in ways he could not have imagined.

There is a theory that Gerard Mercator's son, Rumold, came to Kassel to tutor Wilhelm in maths and how to use instruments during this period. This is an interesting idea, but it is unlikely. Rumold was born in 1541, so would have been just ten or eleven years old in the early 1550s. There is, however, a later connection between Hesse and the Mercators. In 1587, Wilhelm commissioned Arnold, Rumold's brother, to make a survey of his lands but he died of pneumonia before the project was complete. By this time, the great geographer and his family had been living in Duisburg for several decades, just a few days' ride to the east from Hesse.

Between 1559 and 1561 Wilhelm studied with Andreas Schöner, son of Johann, the famous Nuremberg mathematician. Schöner spent two years at Kassel with Wilhelm. They worked on astronomy and made observations together, recording the coordinates of several stars at this time. Wilhelm's enthusiasm for instruments crystallised when he acquired Peter Apian's extraordinary book *Astronomicum Caesareum* (Astronomy of the Caesars, 1540), which featured stunning fold-out paper instruments. A kind of erudite astronomical pop-up book, it was commissioned by Charles V to make celestial observation easier to learn and understand, providing a practical way into this recondite subject. Manufacture was a challenge: the design took Apian eight years, and producing it was extremely complicated and expensive – only about one hundred copies were made. It contains a fold-out torquetum that can be used to make observations. This complex, multi-layered device can measure three separate coordinate systems at the same time using its various frames, scales and discs. Of the thirty-six woodcuts in *Astronomicum Caesareum*, twenty-one are volvelles – rotating paper wheels representing the movements of various celestial bodies. Wilhelm was entranced and made his own instrument using sheets of copper moved by gears. Schöner and Wilhelm worked on this project together – their notes survive in a copy of *Astronomicum Caesareum* in the city archives.[4]

In 1558, Wilhelm observed a comet with a wooden torquetum. Once he had worked out the latitude and value of the ecliptic, he could set these on the torquetum before taking the measurements.

There was also a quadrant which Andreas Schöner described as being 'five feet in height'. Wilhelm used it to observe the meridian altitude of several stars and the moon, the meridian of Kassel and the altitude of the pole.

While there is plenty of information about the devices used at Kassel, we know much less about where they were stored and used. In around 1560, Wilhelm had two three-storey towers built on the south-western and south-eastern sides of the palace, with roof terraces or balconies (*Altanen*) where he stood to make observations. These may have been situated above the kitchens on one side, and a bakehouse on the other; some sources claim they had revolving domes or floors so the instruments could be pointed at any part of the sky. Whether or not this qualifies as the first European observatory in modern times is up for debate; nevertheless, it was where Wilhelm and his astronomers carried out most of their work.

The problem of storing large, unwieldy instruments was an important consideration. Keeping them outside in all weathers was impractical (even the ones made entirely of metal – brass is not water-resistant), so they were probably kept in an adjoining room and moved outside before use. In a letter written in 1586, Wilhelm's court mathematician mentions taking a brass quadrant from the balcony into Landgrave Philip's chamber (presumably his father's old bedroom). This gives us an insight into the everyday workings of the observatory and begs the question of how they moved these large, heavy devices. It is possible they were on stands with wheels or that the instrument could be detached from the stand and taken inside. They could also have had covers to protect them from the elements.

In the early 1560s, Wilhelm began to design his own astronomical instrument, a beautiful clock that replicated the movements of the heavens in all their glorious irregularity, something they achieved by making the spacing between the teeth of the drive wheels unequal. The stars are delicately engraved on a celestial globe at the top which rotates from east to west, while the sun and planets have their own discs mounted on the four sides of the clock, a perfect representation of the geocentric model of the universe. The 'Wilhelmsuhr' (Wilhelm's clock) was made in Marburg, the second city of Hesse, in the newly founded royal workshop. Eberhard Baldewein was in charge, assisted by a watch-maker named Hans Bucher and several court carpenters, using Wilhelm's drawings and astronomical data he had gathered himself; not all Renaissance rulers with similar interests were will-ing to sit at the bench in the workshop and toil beside their crafts-men to produce something of breathtaking beauty and technical genius. The project was a huge success, and made Wilhelm famous. The French humanist Peter Ramus wrote, 'The Germans say that Eberhart [Baldewein] who was once a tailor [sartore] was made into a singularly fine artificer in astronomical matters by the Landgraf of Hesse who is endowed with Democratean clever-ness.'[5] This was common in the early years of the instrument making trade – people began as goldsmiths or some other skilled craft and transferred their expertise under the direction of their patron. A couple of years later, they made another astronomical clock in the workshop for Wilhelm's friend, August, Elector of Saxony, which can still be seen today in a museum in Dresden.

August was a prominent member of the circle of admiring rulers who looked to Wilhelm and Kassel for inspiration in almost every area of their lives. His interests were mapping and geogra-phy, and he used tools like odometers and pedometers himself to measure his domains. Like Wilhelm, he oversaw collaboration between the craftsmen and mathematicians he appointed, helping them to design new instruments for his scientific projects. The workshop August founded in Dresden continued well into the seventeenth century, under the auspices of the Trechsler family who he had initially employed as armourers. He and Wilhelm were lifelong friends, sharing ideas and information on a wide range of subjects.

Otto-Heinrich of the Palatinate (1502–1559) was a generation older than Wilhelm and August, an inspirational figure who designed several sundials himself and commissioned other mathematical instruments, working closely with scholars at the University of Tübingen. In 1554 he commissioned an astronomical clock from Philip Immser, professor of mathematics at Tübingen. Despite collaborating with astronomers and a clockmaker from Heidelberg, Immser struggled to produce a device of sufficient accuracy and was driven to despair by the project. Otto-Heinrich died soon after and the clock was never completed. Making complex instruments was a huge challenge and at this stage, very few people had the capability. It is no wonder the Wilhelmsuhr was greeted with such admiration.

Julian of Braunschweig was another of Wilhelm's correspondents and a keen patron of the mathematical sciences. At his court in Wolfenbüttel, he employed Johann Tiele, a universal craftsman who worked simultaneously as a surveyor, cabinetmaker, carpenter, architect, armourer, cooper and, of course, an instrument maker. At this early stage of technical development, court employees were expected to turn their hand to anything that was required, learning new skills as they went and led by the interests of their patrons. This circle of scholar princes also included Wilhelm's three brothers; together they created the conditions for technological advancement and vital investigation into the natural world that transformed intellectual life in northern Europe.

Mechanical devices like astronomical clocks had various functions. They were objects of beauty and wonder to be proudly displayed in princely collections, and moving replicas of the heavens. If we think about them in similar terms to apps like SkySafari and Star Walk, which highlight and identify the constellations and planets as you sweep your phone across the night sky, we can get some idea of how powerful these sixteenth-century representations of the universe were. For many people, the complex mathematical models in many astronomical textbooks were too difficult to comprehend. Planetaria brought this information to life, and helped to change the study of astronomy into an active, practical three-dimensional endeavour based on observation. They also made it possible to predict future celestial movements, something that had many applications in astrology.

By the early 1560s, the Marburg workshop was also busy building small, spring-driven 'minute-clocks' to be used in the Kassel observatory alongside a quadrant and a sextant, rather than the torquetum, to measure star positions by time rather than angles. Wilhelm believed they were more accurate but, initially at least, they weren't. When compared with contemporary computer-generated data, the earlier torquetum measurements are, in fact, closer because the clocks were not yet precise enough. However, the move to time-based observations led by the Kassel astronomers was important, one that would soon produce better quality results and open new avenues of possibility. Wilhelm also began calculating coordinates using spherical geometry instead of a globe. As always, improved accuracy was the goal.

In 1567 the old landgrave died, and the lands of Hesse were divided into four, one part for each of the sons. Wilhelm inherited Kassel and the lion's share, Ludwig was given the city of Marburg and the surrounding area, Philip got Reinfels and

George Darmstadt. The four areas were to be ruled in cooperation with one another through shared institutions and courts. A
year earlier, Wilhelm had married Princess Sabine of Württemberg,
whose sister Hedwig was already the wife of his younger brother
Ludwig. They had eleven children together and seem to have
enjoyed a happy marriage; he was devastated when she died in
1581. He also continued the work of modernising the old Gothic
castle in Kassel and the city itself, starting with the all-important
fortifications. He replaced the old circular bastions with angled
ones and renewed the gates with the help of expert military engineers. He spared no expense with the creation of his parents'
enormous tomb in the church of St Martin, sending for masons
all the way from Brussels and Antwerp to make the sculptures
that decorate it and celebrate Philip's pivotal role in the establishment of Lutheranism.

One of Wilhelm's greatest achievements as ruler was a huge
political project later called the 'Economic State' (*Ökonomische
Staat*), which involved collecting data and producing a statistical
record for the whole region and its finances. Using surveys
Wilhelm personally created, Kassel civil servants determined
population numbers, resources and property; they recorded the
rights and dominion of each settlement and estate, drawing up
detailed maps showing transport links, mines, saltworks and
forestry. This large-scale mathematisation of bureaucracy was
visionary and remarkable for the time; it enabled Wilhelm to rule
much more effectively. He took his role as landgrave extremely
seriously, taking personal responsibility for ensuring enough food
was put by each year so that none of his subjects would starve,
having farmland assessed and surveyed to ensure optimum
productivity, rivers mapped and measured, water supplies
reviewed – no corner of his realm was overlooked. In his capital

city, home to around 5,000 people in this period, he expanded the salt industry, built new slaughterhouses and increased the size of grain stores, replacing many of the old half-timbered buildings with stone.

This love of order and information based on careful observation ran through Wilhelm's entire reign. He is best known as an astronomer, but he was also a committed botanist who collected and grew flora from all over the world. One list he made includes a smoke bush, Indian marshcress, strawberry trees, epimedium (which was native to China), Tino lilies from Asia, grey-leaved rock roses in pink and white, 'cauli fiori' – presumably a type of cauliflower – and Terebinth trees from the Mediterranean region. He harnessed a wide network of scholars, notably two celebrated naturalists, Joachim Camerarius the Younger and Carolus Clusius. They corresponded regularly and visited Kassel to consult on the expanding gardens. Wilhelm does not seem to have ventured much outside his own familial lands, instead accessing information and specimens from further afield through the network he built up and by sending agents to act on his behalf. Clusius was chief among these. Yet another alumnus of Louvain University, he travelled widely around central Europe before setting off to study the botany of Spain and Portugal in 1564 with Anton Fugger, a young member of the trading dynasty. The emperor Maximilian summoned him to oversee the imperial gardens; he also visited England and brought plants back from Crete for his noble patrons.

Camerarius, on the other hand, was a physician from Nuremberg, where he created his own garden of medicinal plants. He was exceptionally well travelled and had spent time studying in Italy, at Padua and Bologna universities, building up a large network of correspondents with whom he shared information. He often passed these letters on to his patron in Kassel, providing the

German prince with useful connections in Italy. Wilhelm was directly in touch with some notable Italians; he wrote that the Duke of Florence (Ferdinando I de' Medici) 'has sent us some beautiful plants and seeds among which are: Fiscus Indica [Barbary Fig or Prickly Pear], Dactyli, Cameripbes, Balustrian [Pomegranate], Lines, Iunijperus Syriaca [Syrian Juniper], Jujubae Nepa [a type of plum from Central Asia] and others which we never thought would come into this country'.[6]

Camerarius was indispensable to Wilhelm, acting as an agent who sourced plants and responsible for sending his letters on to recipients in Italy, and vice versa. In the preface to *Hortus Medicus et Philosophicus* (1588), Camerarius makes it clear that this was entirely mutual, writing,

> the most illustrious and most excellent Prince, the Landgraf Wilhelm of Hesse…holds a singular love of all the arts…[he] did not hesitate to provide for himself an exact knowledge of plants and not only cultivates in his garden a number of rare and select plants, which Germany has never had before… Indeed, I particularly am indebted to him and I openly admit having received not a few specimens, brought to me by his highness before all others.[7]

Like so much of sixteenth-century culture, the fashion for gardening was brought north from Italy by students and travellers, first to the Habsburg court from where it spread to the German principalities, the Low Countries, Britain and the Scandinavian kingdoms. The Fuggers had one of the earliest, a huge, enclosed garden at their house in Augsburg, filled with rare and unusual plants they had imported through their vast trading network. Charles V visited in 1539 and was struck with a combination of awe and envy. Before long the imperial gardens in Vienna were planted with a wide variety of new specimens brought back from the Americas, and much energy was poured into discovering the best conditions for growing them – the unpredictable European climate often caused problems. Tropical plants required constant warmth and protection from the frost, and soon no princely garden was complete without ornate, heated glasshouses for the cultivation of delicate specimens such as melons and pineapples. As any gardener will tell you, growing non-indigenous species has its challenges, and Wilhelm's concerns about his orange trees, planted on the floodplain and subject to seasonal inundations, are as understandable as they are familiar. We can imagine these sun-loving, Mediterranean plants yearning for the banks of the Guadalquivir rather than the Fulda.

For many, these pleasure gardens were simply an expression of their wealth and sophistication, a status symbol that trumpeted their cultural credentials. This was true for Wilhelm, but he was also motivated by the desire for knowledge of the natural world. Each flowerbed and plantation was considered with characteristic focus and forensic attention; he approached gardening in a way we would now describe as scientific. Every plant was meticulously described and studied, ordered and identified, while the best procedures and conditions for cultivation were tested and noted. Ever the pragmatist, Wilhelm was also an early adopter of the ultimate meal on a budget, the humble baked potato, although they were neither cheap nor ubiquitous in those days. He was one of the first in Europe to grow them and sent some, with instructions on how to cook them, to his old friend August of Saxony.

Just as the concept of the pleasure garden arrived from Italy, so did many of the plants themselves. Wilhelm sent his gardener to purchase specimens in Venice; he ordered myrtle, lemon and pomegranate trees directly from Tuscany through the Thurisan trading family of Nuremberg, who acted as his agents. He enthusiastically shared plants and tips with his brothers, his numerous aunts, friends and many fellow sovereigns, sending Frederick II of Denmark a list of forty seeds and twelve plants for the new garden he was planning. The Hessian archives are full of letters asking Wilhelm for advice, and he seems to have acted as horticultural consultant to most of northern Europe. The Kassel gardens were a huge source of inspiration and plants for the many people who visited, just as so many other aspects of his court were emulated as a model of how things should be done.

Wilhelm's wife Sabina shared this love of botany. Like Philip I in Hesse, her father Duke Christoph von Württemberg had established Lutheranism in his lands and was a keen follower of Melanchthon, so she would have grown up in a similar atmosphere to Wilhelm. Before the wedding in Marburg in February 1566 she was warned that her husband-to-be was 'a strange stargazer', but her sister Hedwig, already married to Wilhelm's brother Ludwig, wrote letters reassuring her everything would be alright. Girls in this period were taught from a young age that being a good wife meant accepting your husband's peculiarities, no matter what they were. In the end, this was not too difficult, and the couple grew to love each other dearly and had many shared interests.

Sabina gazes calmly out at us from the double portrait painted of her and her husband in 1577. Her dark eyebrows frame a serious, contemplative face, her hands are neatly folded across her stomach, modest white ruffs at her wrists and neck. She wears a lush dark gown over a crimson robe with long sleeves, and several ornate chains are looped around her neck. Her hair is hidden by a heavily jewelled cap, on top of which she wears a black hat embroidered with gold thread. She looks sensible, thoughtful, intelligent.

Wilhelm is equally sober, in typical Lutheran dress, dark fabric illuminated by two small gold chains and a thin belt. His ruff is similarly plain, open at the chin where his beard forks down. Between them is a view of their beloved garden, symmetrically laid out with trellises, paths and flowerbeds, decorative structures for climbing plants and a neat hedge around the outside, channels of the Fulda encircling it under a bruised sky. The couple stand on a balcony or platform, surrounded by astronomical instruments; this is the only surviving image of the Kassel observatory. Behind Wilhelm's left shoulder is a celestial globe, its brass armature gleaming, as it still does today in the Alte Meister Museum in Kassel, also the location of the two portraits. On his other side, two men can just be seen, holding a quadrant. One of them is Eberhard Baldewein; the other may well be Tycho Brahe, who had recently visited Kassel. Sabina stands next to a black torquetum that reaches to her elbow. Behind it is a red azimuthal quadrant.

Wilhelm had a 'tasting house' built for Sabina in the little pleasure palace where she distilled medicines using herbs she grew in the garden, skills she learned from her mother, Anna Maria, during her childhood at the court in Stuttgart. The closest we can get to Sabina is a book of medical recipes now in Heidelberg University Library.[8] It was handwritten by several people (but not Sabina); there are some notes in the margin stating that some of the recipes are hers. There is also a short chapter on her in a modern history book about German noblewomen of the period.[9] Her life was defined by pregnancy and her devastation over the loss of several of her children. One night in the summer of 1580, Sabina and Wilhelm were staying at their hunting lodge in Haydau. They went out to gaze up at the stars together and she had a premonition of her death. In February the following year, she gave birth to her eleventh child, Juliane, who only lived for two days.

Sabina could not bear the grief and died herself a few months later, leaving Wilhelm inconsolable.

Sabina treated the people of Kassel, administering the medicines she made for free and, in her will, she left provision for this care to continue, which it did, right up until 1936. It is difficult to find direct evidence of Sabina's involvement in the gardens, but Wilhelm carried on a lively correspondence with his brother Ludwig and Sabina's sister Hedwig, both passionate botanists at their palace in Marburg. In one letter, Hedwig promises to send on to Kassel any unusual plant specimens she is able to get hold of. The effort to record and order plant species was a primary concern for both Wilhelm and Ludwig. When a plant was too delicate to be transported between the two courts, an artist would make a detailed botanical study that could be used to assess and identify the specimens. Ludwig consulted regularly with the medical professors at the University of Marburg – pharmacology and distilling herbal medicines was a key motivator behind the identification programme. Wilhelm also collected the most up-to-date books on botany by the likes of Dodanus, Lobelius and Clusius for his library at Kassel, all published by our old acquaintance Christoph Plantin in Antwerp. Wilhelm commissioned extra drawings for his copy of Lobelius' *Plantarum seu Stirpium Historia* (which already contained 1,445 plates) and in 1581 he helped Camerarius to purchase the library of the acclaimed naturalist Conrad Gesner. Time and again, Wilhelm proved himself to be much more than just a cultivated Renaissance gentleman; he was an integral player in the movement to re-establish the study of botany and horticulture on a rigorous, humanist basis.

Wilhelm may have inherited the lion's share of his father's estate, but he did not get Eberhart Baldewein, who remained in Marburg at the workshop and took on the role of *Beaumeister* (master builder) for Ludwig. The brothers bickered gently over who should have the greatest claim on his time, and Wilhelm continued to commission him 'to employ his talents in the construction of buildings, various artefacts, instruments and in other such things of which he has an understanding'.[10] This put the former tailor under a huge amount of pressure; without a clearly defined role he found himself responsible for several, effectively full-time, jobs.

Baldewein's talents and the skills he had built up were not commonly available, as Wilhelm discovered when he brought Hans Bucher to oversee the new Kassel workshop in 1574. Although skilled, he had 'no head for invention' and by 1579, he had been replaced by a new craftsman, Jost Bürgi.[11] Wilhelm was more fortunate with his new appointment. Bürgi was Swiss and seems to have been blessed with many of the characteristics that country is known for today. He was exceptionally skilled at clock-making, a strong mathematician, precise, efficient, modest, discreet and hard-working; he also invented several ground-breaking pieces of technology. Johannes Kepler, who worked with him in Prague, said,

> Jost Bürgi, the maker of moving models, who, although he knows no languages, yet in the science and consideration of mathematics easily surpasses many professors of these. He possesses a dexterity so peculiar to him that a later age may celebrate him as a true leader in this genre, no less than Dürer in painting, whose fame grows imperceptibly like a tree with time.[12]

The landgrave took no chances with his new employee, drawing up a detailed contract that explained exactly what was expected of him. He was to live at court and oversee the palace clocks and instruments; when commissioned, he would also be responsible for making new devices. In return, he would be paid thirty gulden a year and 'we shall provide him with the usual sort of courtly attire and board at the court with other of our mastercraftsmen; and with this we shall also arrange for his room and lodging...and for the necessary wood and coal for his heat and that which is necessary for his work.'[13] This was an important step towards formalising the emerging profession of instrument maker, another area where Wilhelm led and others followed. He may well have been put in touch with Bürgi through his old contacts in Strasbourg, where the clockmaker had been helping to build the cathedral's astronomical clock, several decades after Wilhelm himself had been so inspired by it.

Timekeeping and the construction of accurate clocks became the defining characteristic of Wilhelm's astronomical work. He pioneered the use of time in celestial observation, a crucial innovation that led to improvements in the data produced at Kassel. Like many astronomers before him, Wilhelm noticed discrepancies between the positions of the fixed stars and what he could see in the night sky. As early as 1558, he resolved to produce a new star catalogue based on his own observations to correct the errors that had plagued astronomy for a millennium and a half in the tables listing star positions recorded by Hipparchus and Ptolemy in antiquity. As we have seen, the need for new star tables was a key tenet of Regiomontanus' reform programme, as it had been for Ulugh Beg in the early fifteenth century. There is no evidence that Wilhelm was aware of Ulugh Beg's observatory in Samarkand, however, he was directly linked to Regiomontanus by his relationship with Andreas Schöner whose father Johannes had published several of Regiomontanus' works and lived in Nuremberg for a time. He would certainly have seen and known about the balcony

Walther had built onto the house in the first years of the century, where he used his instruments to make observations. It is entirely possible that Schöner told Wilhelm about the observation balcony and so helped inspire him to create his own in Kassel.

Wilhelm made great use of Regiomontanus' treatise on instruments, which Schöner published in 1544. He soon realised that accurate clockwork would be essential to producing a new star catalogue. Bürgi set to work and by the early 1580s there were three clocks at Kassel that were used alongside the instruments to measure celestial positions. Crucially, one clock had 'three hands, which indicate not only hours and minutes but also individual seconds', thanks to Bürgi's invention of the 'cross beat' escapement.[14] This ushered in a whole new realm of precision that Wilhelm was keen to share with his fellow scholars. When there was a discrepancy in their two sets of data, he sent Caspar Peucer, professor of mathematics at the University of Wittenberg and court physician to August of Saxony 'a beating clock together with a quadrant. And such a clock might the Wittenbergers well employ, for then with the extra truth and exact knowledge of the middle of the sky, parallax cannot be detected.'[15] Parallax is the apparent change in location of an object when viewed from different positions, for example a person's eyes, or two different places on earth. People who believed in the Copernican system were aware that the earth was moving constantly so parallax should be expected. The greater the distance of an object, the smaller the parallax; this led astronomers to eventually accept that the stars and planets are much further away, and therefore the cosmos is far larger, than previously believed. Wilhelm figured that to be able to produce truly comparable measurements, they needed to use the same equipment. In doing so, he helped establish the parameters of working collaboratively from different locations – something that is a crucial feature of modern science, and in particular, astronomy.

When Wilhelm became landgrave, he was no longer free to stay up all hours on his observatory platforms gazing at the stars. The new catalogue stalled, as other projects monopolised his attention. Then, in November 1572, something extraordinary happened in the night sky – an unbelievably bright new star appeared in the constellation of Cassiopeia. If you had to pick the most important celestial moment in the sixteenth century, this would be it. This star, this supernova, was, in fact, not new at all. It was incredibly old, and it was dying. Its sudden brilliance was caused by twin stars combining as hydrogen atoms fused, turning one into a white dwarf. This dense burning heart of the old star, spinning wildly, pulled in and subsumed the other, combining them into one giant entity, a stellar merger. When their combined mass reached a critical point, the nuclear heat and force inside caused a massive explosion that beamed across the cosmos, visible to everyone on earth, the brightest star in the night sky. This light faded gradually over the next few months – by summer the following year Tycho's Supernova, as it came to be called, was no longer visible to the naked eye. In the 1990s its remnants were photographed by a NASA space telescope, the Chandra X-ray Observatory, still glowing in a far corner of the cosmos, 450 years later.

Astronomers all over the world, from China to Sweden and beyond, saw it and marvelled. Dee in his garden, Brahe with his uncle, Tadeas Hajek in Prague, ad-Din in Istanbul and the landgrave on his balcony – they all gazed up in wonder. This was one of those rare moments when everything begins to change. At this point, astronomers had only recently begun to expand their investigation of the universe beyond hypothetical mathematical models. Aristotle still dominated conceptions of the physical composition and organisation of the universe, in the realm of philosophy that was based on conjecture rather than experience. His theory was that the earth is nested inside a series of crystalline spheres on which the stars and planets circle in the perfect, unchanging heavenly realm. People already had serious doubts about this idea, but the appearance of a new star in this realm shattered it once and for all. How could the heavens be unvarying if a new star could appear out of the blue? This was an astounding occurrence and it had a transformative effect on the study of astronomy, inspiring stargazers, Wilhelm included, to redouble their efforts to study and comprehend the mysteries above them.

Within days messengers were racing across the continent to deliver letters as astronomers rushed to compare observations and theories with one another. Wilhelm's interest in the stars was reignited, so much so that he refused to leave the observatory when a fire broke out in the palace. August of Saxony first alerted Wilhelm to the appearance of the supernova. He had observed it alongside Caspar Peucer, who worked out the star's distance from earth and was astonished by the result. It appeared to 'just touch the sphere of Venus and transcends the orbs of the moon and Mercury'. In other words, it was deep inside Aristotle's sublime, unchanging celestial realm. He reported that this discovery 'disturbs our reasoning' and 'pulls my soul in different directions'.[16] Peucer's reaction shows the profound challenges posed by this type of paradigm change; indeed, he continued to believe in the idea that the celestial zone was unchanging, explaining that the new star was an isolated anomaly and therefore not enough to overturn it. He certainly wasn't alone; it took several decades (and the invention of the telescope) for the astronomical community at large to accept these new theories.

As always, Wilhelm's reaction was diligent and methodical. He was not going to come to any conclusions before comparing data on the new star with as many others as he could. He wrote to his wife's brother, Ludwig of Württemberg, who put him in touch with Peter Apian's son Philip, at the University of Tübingen, and he also received data collected by the mathematician Cyprianus Leovitius in the Palatinate. By the time of the appearance of the supernova, there were at least two more quadrants in the Kassel observatory, both made of brass. As time went on, clocks became increasingly important to the observation programme, used to time the movement of celestial bodies past certain points and then calculate the angles from there. Bernhard Walther had used these methods in Nuremberg, albeit with far less precise timekeeping, and Brahe had the same issue on Hven – his clocks were not sufficiently exact to use in observation. In the seventeenth and eighteenth centuries, as clocks became increasingly sophisticated, they would become central to astronomical data collection and today modern astrometric satellites like Gaia and Hipparchus still use timing to determine position in space.

In 1575, Tycho Brahe arrived in Kassel. Given the amount these two men shared, it is surprising he had not visited sooner on one of his previous journeys around Europe. This time, Brahe was on his way to northern Italy, to Venice and Padua, searching for skilled craftsmen to tempt back to Denmark to help build King Frederick II's splendid castle at Elsinore. By this point he had made several journeys around Germany as a student, to Leipzig, Wittenberg, Rostock and Augsburg, even reaching Basel in Switzerland in 1568. However, this was his first time travelling in an official capacity as a nobleman, with a baggage train, a burgeoning ego and everything else that entailed. It was also his first time abroad as a renowned, respected, professional astronomer. He arrived at Kassel in spring 1575 and, as befitted his noble status, was welcomed by the landgrave himself. With affairs of state weighing so heavily on him, Wilhelm was frustrated by how little time he had to devote to astronomy. It is surprising that he had still not yet employed a court mathematician to make regular observations for him. Tycho's arrival gave him the excuse to cancel his official duties and spend several happy days gazing heavenwards. The two men had so much to talk about; instruments were at the top of the list. We can imagine them deep in conversation in the palace, examining Wilhelm's new quadrant and making observations together on the balconies.

They discussed the new star of 1572 and compared data, agreeing that it was located further into space than the moon and must, therefore, disprove Aristotle's claim that the heavens were unchanging. Tycho encouraged Wilhelm to step up his long-term project of creating a new star catalogue, and possibly put him in touch with a scholar named Hans Buch, who joined the court of Kassel later that year as court astronomer. His appointment was vital in getting astronomy at Wilhelm's court going again and during their conversations, Tycho and Wilhelm talked about which issues they considered to be most pressing and which directions their research should take. Wilhelm gave Tycho the data he had collected during his twenty or so years of observation, giving him the benefit of his increased age and experience (he was fifteen years older), as well as highlighting the value of having several years' worth of observations to work with. The two men never met in person again, but they corresponded for many years and continued a fertile scientific partnership.

There can be little doubt that Kassel was a huge source of inspiration for Tycho. Here was a beautiful house, home to Wilhelm's large family and also to his books, instruments and collections, a true stargazer's palace. As we have seen, Wilhelm's staff set up his equipment on the balconies to give a clear view of the night sky, moving them to the other side of the building when necessary. It is possible that this arrangement made Tycho aware of the importance of having dedicated observatory spaces where equipment could live and be able to access any part of the firmament he wished. As we shall see in the following chapter, he ended up building two observatories at Uraniborg, one in the main building with a retractable roof, and a separate one underground called the Stjerneborg, which benefitted from darkness and far more stability than anywhere above ground.

Tragically, just a few days into Tycho's visit, Sabina and Wilhelm's baby daughter Sidonie fell seriously ill and died. The court went into mourning and Tycho packed his bags, cutting his visit short, but there were still positive consequences for both men. Wilhelm was rejuvenated by the younger man's enthusiasm and followed his advice of employing a court astronomer so that the star catalogue project could progress. In turn, Wilhelm urged Frederick II of Denmark to support Tycho and ensure he settled at home, pointing out the prestige and advantages this would bring to his country.

In 1577 the heavens sprang another surprise – a dramatic comet that was visible for two and a half months. This gave ample opportunity for multiple observations and detailed measurements of the comet's journey across the skies. Like the supernova of 1572, it was clearly in the super-lunar realm, crashing through Aristotle's crystal spheres as it moved through space. For some leading astronomers, this was the nail in the coffin for both Ptolemy's heliocentric universe and Aristotle's unchanging one. Reaction to these two events reverberated through an unprecedented wave of letters, books and pamphlets as scholars debated what they presaged, and what they meant for how people understood the cosmos. Almost every text contained astrological predictions and a discussion of what the star and the comet foretold; for most people, this was the most important aspect of celestial activity – the effect it would have on events on earth.

Wilhelm's attitude towards astrology was, as we would expect from someone with his intellect and character, nuanced. He did not believe in direct, 'divinatory' astrology, that the heavens dictated human behaviour. However, he did allow for the heavens causing physical effects that influenced natural objects, and the idea that these could be understood as divine. For him, the new star of 1572 was a portent of future events, as was the comet of 1577. He was also interested in planetary conjunctions, when two planets align as viewed from earth. This happens fairly regularly and predicting exactly when they would occur and how long they would last was a major preoccupation in sixteenth-century astronomy because they were believed to presage significant changes or events on earth. Peter Apian discusses their importance in *Astronomicum Caesareum*, reflecting a common theme in astronomical literature. A maximum conjunction of Jupiter and Saturn was coming up in 1583/4 and Wilhelm began work on it as early as 1576. He wrote to August of Saxony, 'This will undoubtedly bring about such a change…that I wish to extract and calculate accurately the times of convergence as well as the times of mean and real conjunction and, therefore, to note the moments of historical change [which coincide] with it.'[17] The problem, as ever, was the lack of accurate star data by which Jupiter and Saturn could be mapped. One solution was to measure the ascension of the sun, which had been done consistently in Kassel for decades and was the most common type of observation carried out there, used to map the stars and planets.

At some point around 1580, Wilhelm appointed Christoph Rothmann as his *mathematicus*. He had studied theology and maths at the University of Wittenberg, but we know little else about him. Wilhelm set him to work measuring star positions for the new catalogue, among other things. In late January 1580, Wilhelm had observed a lunar eclipse and sent his measurements to a maths professor at the University of Marburg, asking him to calculate the predicted time and duration using the Prutenic Tables (based on Copernicus' data). There were huge variations between these and the actual lunar observations. By 1586, he had concluded that the old star tables were useless, and that the only way forward was to continue his entirely new catalogue, begun several decades earlier.

Initially Rothmann was a reluctant participant in this project; he was a tricky character and did not seem to understand why new coordinates were needed. However, he set to work in the observatory. When he had made a set of observations, he compared them to the old tables and was shocked by the results. The discrepancies were so large he initially suspected his instruments were faulty. He re-did the observations several times, always with the same results. Finally, the penny dropped: 'all the learned,' he wrote, 'have complained not without reason concerning the tainted and as yet unimproved stellar catalogue since those same stars, which one up until now believed had been observed very accurately by the ancients, have been found to deviate from the tables by two, three, four or more degrees.'[18]

In 1584, Paul Wittich came to join the Kassel observatory for a while. Known as 'the wandering mathematician', he had spent time at most of the important intellectual centres of northern Europe: Prague, Leipzig, Hven, Frankfurt and Wittenberg, among others. Wittich is one of those shadowy, compelling characters who populate the history of science. No books by him survive; we can only glimpse him occasionally in the correspondence of his peers, who obviously regarded him very highly. Brahe regularly referred to him as 'that most illustrious mathematician of Wrocław', the Scottish mathematician John Craig praised his mathematical abilities, and another fellow scholar wrote, 'Wittich allows neither reason nor advice...to dissuade him from his enthusiasm for travelling. In his very pleasant company for a few days, I have felt myself enflamed for mathematical studies.'[19] A picture emerges of someone with a brilliant mind and a charming manner, whose contribution has been lost in the passage of time. However, he did leave us some intriguing clues. Wittich's love of travel was pivotal; he visited so many places and met a wide range of people, transmitting ideas (his own and those of others) as he went. Having studied in Leipzig and Wittenberg, he visited Tadeas Hajek in Prague and spent time in Altdorf near Nuremberg.

In 1574 he moved on to Frankfurt an der Oder, where he met the aforementioned John Craig. They became friends and discussed maths together. Wittich showed Craig a new method for calculating angles using sines and cosines, and this was likely the very same method that Craig showed John Napier when he got back to Edinburgh – Napier later used it to invent logarithms. The example Wittich showed him was written out on a blank page in his copy of De Revolutionibus, which Craig transcribed exactly in his own copy of the same book. He remained close to Hajek, they corresponded regularly, and when Wittich arrived on Hven in 1580, he had a letter of introduction from the Czech astronomer in his pocket. We will explore this in the next chapter and jump forward here to the time he spent at the Kassel observatory in 1584. Wittich suggested various technical improvements and helped launch a five-year period of intense activity in the observatory. Bürgi implemented his changes, adding transversal lines to the scales of instruments to enable more accurate readings, building a new sextant that could measure distances between the stars and adding slit sights onto the alidade, the measuring scale of the quadrant. All of this increased observational precision. While he was there, Wittich also designed an astrolabe, which is still in the Astronomisch-Physikalisches Kabinett in the Kassel State Museum, a gleaming testament to his brilliance, and the only solid object he left behind.

Rothmann recorded Wittich's suggestions in a manuscript about making astronomical instruments, but, spiky as he was, he could not resist a dig at the Pole, commenting that his observations were unreliable because he had bad eyesight, that he would have been better off sticking to geometry. Wilhelm wrote delightedly to Brahe, 'We have improved our mathematical instruments from the instruction of Paul Wittich so that as before we could observe to two minutes [of arc] we are now able to observe ½, indeed ¼ of a minute.'[20] At this point the atmosphere of cosy collaboration soured dramatically. Brahe was furious that information about his observatory and instruments was being shared without his permission; with characteristic arrogance, he claimed the innovations as his own and attacked Wittich. Hajek, who was in Vienna working for the Holy Roman emperor, Rudolf II, had to weigh in to defend him. Brahe was driven by his anger over this situation for the rest of his life; from then on, he referred to Wittich as 'a certain Wrocław mathematician'. As we will see, vicious competition was as much a feature of sixteenth-century intellectual communities as fruitful collaboration, and Brahe was especially sensitive about the ownership of scientific ideas.

Back in the Kassel observatory, Rothmann, sometimes accompanied by Bürgi or an assistant (in this busy period there were at least three), was spending his nights on the palace balconies, eyes trained on the heavens. On 19 November 1585 he began observing at 7 p.m. and continued throughout the night, while the following evening his watch was shorter, from 9 p.m. until midnight, possibly because of bad weather. The winter of 1585/6 was terrible for astronomy, with clouds and mist obscuring the night sky for weeks, and Wilhelm became agitated that the project was not progressing. There was nothing that could be done. Rothmann often stayed up all night in the hope that the weather would clear: it must have been an arduous, demoralising experience. He reported that in January 1586 it was so cold that the instruments froze onto his skin and he could not slide the alidade. Astronomical observation was not for the faint-hearted – it was (and still is) a demanding occupation requiring great precision in the face of exhaustion, bad weather and frustration.

By 1587 Rothmann was using one of Bürgi's clocks to measure the longitudes of the stars, far more precise than the old readings which relied on angles. The catalogue produced that year, which contained the coordinates of 383 stars, was a great improvement on that of 1567 in its accuracy. Kassel's reputation attracted a constant stream of visitors. The following year Caspar Peucer, finally released from twelve years in prison for being a suspected Calvinist, came to visit Kassel. He later told Tycho how much he had enjoyed using the instruments there; several other scholars shared his enthusiasm, judging by those who later visited because they were keen to see the equipment in the Kassel observatory. It is hard to measure this kind of influence, but we can be sure that it inspired the wider intellectual community. John Dee arrived in 1589. Though we do not know the purpose of his visit, the observatory and library were a clear draw for scholars of his stature. He spent time with Rothmann, who later referred to him as his *amico meo singulari* in a letter to Tycho.

Fortunately, the controversy over Wittich's innovations didn't ruin the relationship between Kassel and Hven. Letters continued to be sent between the two places, data compared, and theories discussed. Rothmann wrote to Tycho describing improvements in his instruments and not long after, one of Tycho's assistants visited to find out what they were. In 1590 Rothmann himself left Kassel to visit Hven. By now he had worked out the coordinates of 400 stars, based on new measurements, for Wilhelm's catalogue, which he left behind in a beautiful, handwritten manuscript. He spent a month there with Tycho but did not return to his position at the court of Wilhelm IV as expected. Instead, he travelled to his hometown of Bernberg and remained there for the rest of his life studying theology.

Given the years of sleepless nights and exposure to the elements on the Kassel balconies, his eyes straining in the darkness, it's perhaps understandable that Rothmann decided to retire to a warm study filled with religious texts. Observation was a young man's game, and he must have been comfortably into middle age by now. Bürgi can only have been relieved; Rothmann was a difficult person, and their relationship was not a happy one. Wilhelm made him court astronomer and in 1591 he became a naturalised citizen of Kassel, marrying a local woman and adopting her much younger orphaned brother, Benjamin Bramer. Bürgi educated the little boy, and he went on to become a great mathematician in the following century.

That same year, Bürgi completed an astronomical clock based on Copernicus' model of the universe, the first of its kind and a bold move at a time when the Church was clamping down on heliocentric theory. Another astronomer named Ursus, who visited

Kassel in 1586, had translated *De Revolutionibus* into German for him because he knew no Latin. Wilhelm made his own position on the Aristotelian conception of the universe clear in 1585: 'The principle of the philosophers is destroyed, that comets should be generated in the upper region of the air under the circle of the moon.'[21]

Rothmann went much further – he was a convinced Copernican and dismissed the possibility of spheres or any distinction between the earth and space. This made him one of the very first to accept the heliocentric universe as a physical reality, not just a theoretical system of planetary movement. Wilhelm, on the other hand, worked within a Ptolemaic universe, but was also open to employing Tycho's or Copernicus' systems for some purposes and was happy to allow his employees intellectual freedom to make up their own minds. The Tychonic system combined elements of both. The sun, moon and the fixed stars revolved around the earth at the centre of the universe as in the Ptolemaic version, while the planets moved around the sun as Copernicus had suggested.

When he was in the workshop rather than the observatory, Bürgi spent a lot of time making mechanical globes using the system of logarithms he had developed himself (independently and probably before Napier's). These were sent to courts all over Europe, consolidating Wilhelm's position as a leading distributor of these devices and further enhancing his already impressive reputation.

Wilhelm died in 1592. Bürgi continued to make observations, mainly of the sun and the planets rather than the stars, so a degree of astronomical work did continue for a few years, before tailing off. Wilhelm's son Moritz kept him on as court astronomer, but his own interests lay elsewhere, primarily in alchemy. After a few years, Bürgi moved to Prague to work for Rudolf II, with whom he had been in touch for a while. The court in Kassel continued to be a beacon of culture and erudition, but the observatory fell out of use and the observations so carefully accumulated over the last half of the sixteenth century lay hidden in the palace, their neat columns unseen by anyone. The main star catalogue manuscript, written out by Rothmann shortly before his departure, is still there today, a stunning illustration of scientific achievement.

Mindful of his former patron's wishes and surely frustrated to think of all his hard work going to waste, Rothmann wrote to Moritz asking him for support in completing and publishing it, but this came to nothing and there is no record of a response. In 1618 another scholar wrote in a similar vein, this time with more success. A book was published soon after, but strangely it was made up of Bürgi's planetary and solar observations, and hardly contained any of the star measurements. The star catalogue was not printed until 1666, around the same time that John Flamsteed, the first English astronomer royal, began discussing Wilhelm's observations and contribution in lectures he gave at Gresham College. The celebrated Polish astronomer Hevelius acknowledged their value by including the Kassel star positions in a book he published in 1690. Before that, some aspects were disseminated through informal networks of correspondence with fellow rulers and mathematicians at the universities of Wittenberg, Marburg and Tübingen, and Tycho Brahe, who published parts of his correspondence with Wilhelm and his astronomers. As we have seen, scholarly letters were often copied and passed around, providing a forum where ideas were shared and discussed, precursors of the scholarly journals that would develop in the following century. Tycho sent them on to Hajek in Prague, disseminating Wilhelm's findings to the circle of scholars at Rudolf II's court. From here

they found their way to Italy, where Galileo was another beneficiary; he considered Wilhelm's observations of the supernova to
be the most accurate available. However, it was Tycho's star catalogue, published in 1602, that triumphed.

This meant that the legacy of the Kassel observatory was fragmented. The star catalogue did not have any influence until the
late seventeenth century and Wilhelm's reputation, unassailable
during his own lifetime, faded quickly. Today he is known only to
people with a specialist interest in the history of science and it is
difficult to find much information on him that isn't in German; I
have only found one academic historian who has written significantly about him in English. Even in Kassel itself, he is not easy to
locate, although this is partly explained by the closure of the
Astronomisch-Physikalisches Kabinett for restoration at the time
of writing. Recently, German historians of science have begun to
recover his reputation, and there is a growing body of technical
articles about his astronomical achievements.

While he was alive, however, Wilhelm's influence was far-reaching and he was praised by scholars across the continent for his
talent. In 1588 the Italian philosopher Giordano Bruno said, in a
lecture at the University of Wittenberg, 'Among the Germans...we
find the great rescuer of [astronomical learning], the great Landgraf
Wilhelm of Hesse. Rather than alienating his senses and the intelligence of his eyes [this prince] learns astronomy by experience.'[22]
His fellow sovereigns looked to him for advice and inspiration in a
wide array of matters; where he led, they followed. When, in 1582,
the papacy reformed the old Julian calendar and sought to impose
the new Gregorian one across Europe, including the Protestant
areas, the German princes immediately looked to Wilhelm for
guidance. Being interested in the curiosities of nature was a key
characteristic of Renaissance nobleman, but Wilhelm went far

beyond this. The court at Kassel exhibited many of the qualities we now think of as scientific, although the term did not exist in that sense for several centuries. By observing natural phenomena systematically, making and using instruments, sharing practices and equipment so that data was comparable and working collaboratively with others, Wilhelm and his network prefigured the developments in science that we now take for granted. Similarly, his attitude was always pragmatic and slightly sceptical. He was only interested in alchemy for its practical use and believed that God alone had the power to transmute other metals into gold or silver, in stark contrast to almost every other monarch at the time. He was vociferous in this regard, frequently warning his fellow rulers about the dangers of this type of alchemy. In a letter written to Wilhelm's son Moritz, John Dee recalled how Wilhelm managed to keep 'sophists promising mountains of gold at arm's length'.[23]

Much of the work going on in Kassel under Wilhelm's aegis anticipated the 'new' science of Francis Bacon and René Descartes of the following century. The court at Kassel abounded with collections of curiosities, botanical gardens, an impressive library with books on 'every art', and a glass works where high-quality vessels were produced using ash imported from England via Antwerp and used coal – an improved method which Wilhelm shared with Frederick II of Denmark. There was also a focus on metallurgy and developing improved methods of extracting metals, part of the general trend of exploiting the natural world for profit in Europe. Finally, there were Wilhelm's workshops, where state-of-the-art instruments were developed and manufactured by experts employed for that purpose – in this regard, the landgrave's influence was significant. Courts across Germany and beyond would follow his example, professionalising instrument makers and mathematical scholars all over northern Europe.

During the sixteenth century, the link between intellectual accomplishment and royal prestige became increasingly powerful, thanks in part to Wilhelm's court at Kassel. For medieval rulers, strength in battle and skill at hunting were the most keenly celebrated attributes, but Renaissance princes also sought to fill their courts with scholars, artists and craftsmen, people who could make them glitter with intellectual brilliance, beauty and ingenuity. In the Islamic world, caliphs had been boasting of these kinds of wonders for centuries, and Europeans looked south and eastwards in terror and admiration. Some of these rulers had built observatories filled with specialist equipment for observing the stars: astronomy was a popular area of study. Wilhelm and his court at Kassel were part of this tradition, a stepping stone on the journey to the permanent, institutional observatories that were founded in the following centuries. By making celestial observation a consistent, long-term activity, he demonstrated the importance of accurate data collected over significant periods of time for the development of astronomy. As we will see in the next chapter, Kassel provided a blueprint for another influential palace observatory and the interaction between them would have far-reaching consequences for the evolution of science.

5

HVEN

ISLAND OF URANIA

His greatest pleasure was research, science was his wealth, nobility his glory, in religion he found his delight. He was free from affected manners and from hypocrisy, and he always called a scoundrel a scoundrel. Hence the hatred he incurred.

Johannes Jessenius, *Funeral Oration for Tycho Brahe*
(Prague, 1601)

The little island of Hven lies between Denmark and Sweden in a narrow, but lucrative, stretch of water called the Øresund – for many years, every ship sailing in either direction had to pay tax to the Danish treasury. Hven has high cliffs, yellowish in colour and unusual in this part of the world. To get there today, you take the ferry from Landskrona on the southern Swedish coast. Ven, as the Swedes call it, belongs to them now, but in the sixteenth century it was part of the huge territories ruled by the Danish king, along with much of southern Sweden 'Scania' and Norway. In those days it was as closely connected to the mainland to the west as to the east, and boats regularly crossed between the two, often stopping off to deliver supplies or visitors. Today they are linked by the longest bridge in Europe and a tunnel that meet on an artificial island in the middle of the strait. In the late sixteenth century, however, Hven was much like any other Scandinavian island, populated by a few hundred peasants who tilled their neat fields, worshipped in the little white church of St Ibbs and enjoyed a degree of autonomy and independence. That all changed in 1576 when the Danish king, Frederick II, awarded the island, and all its inhabitants, to a nobleman called Tycho Brahe. Over the following decades Tycho built an astonishing observatory, a beautiful stargazer's palace, on the island and made the name of Hven famous all over Europe.

Denmark is a watery land. Water surrounds it, encircling its many islands and washing at their sandy shores. Water divides it, carving out the topography, leaving just the highest points visible above the waves: a floating country only just managing to keep its head above sea level, the flat, grassy land fringed with salt marshes where lambs graze. Small streams flow into its innumerable lakes, of all shapes and sizes, but they are not filled by water that has tumbled down mountain slopes, for there are no mountains. This water has seeped upwards through the soil, or slipped, imperceptibly, over the low-lying ground, inundating it with dark pools, channels and swamps. In the cold, murky winters all this liquid freezes – sometimes even the sea grows sluggish and solidifies, jagged with ice. There is no drama in this landscape. The hills are gentle and rolling, the soil is fertile; fields of potatoes, red cabbages, strawberries and asparagus undulate into ancient forests of oak and ash. Lilac grows wild in the hedgerows, cows grow fat in the pastures, tidy rows of trees line the roads, and whitewashed farmhouses are arranged around cobbled courtyards – everything is as it should be. Tycho Brahe grew up in this blessed land, drinking rhubarb cordial, feasting on raspberries with thick yellow cream in the summer, and bacon, chestnuts and silver herrings on dark bread in the winter months.

Tycho was born into the highest echelons of Danish society, with not one but several silver spoons in his mouth. All four of his great-grandfathers and both his grandfathers had served on the ruling council that supported the crown, as would his brothers and brothers-in-law. But Tycho was different. He was not remotely interested in the things this life had to offer – governing the country, running vast estates, or indulging in the endless social whirl. All he wanted to do was gaze at the heavens, design instruments, and write books. In order to fulfil his dreams, he had begun to carve out an atypical path for himself in a society governed by rigid structures and defined roles. Tycho did not have much time for these rules; he was comfortable with anyone who shared his passions, regardless of their background, which he first experienced as a teenager boarding with his university professor and spending time with his fellow students in the Latin quarter of Copenhagen.

In the 1570s Tycho was back in the capital lecturing at the university himself. As the first nobleman to take up a teaching role of this kind, he found himself moving between the respectable, well-educated middle class of the city and the elite environment of the court. While this was not without its challenges, Tycho was treading his own path and at ease in both of these worlds, happily traversing between the two. He was determined to live free from the constraints of noble life and chose a wife who would guarantee this. Aristocrats invariably married one another, and most unions were basically financial transactions between wealthy families, arranged by the couple's parents. While he was living with his Uncle Steen at Herrevad Abbey (in what is now southern Sweden), Tycho fell for a girl called Kirsten Barbara Jørgensdatter, who may have been the daughter of the local vicar, far below him on the social scale. The only possible union they could form was a morganatic marriage – an arrangement that allowed them to spend their lives together, but barred Kirsten from fulfilling any official role as his wife and their children from inheriting anything, including the Brahe name. This type of relationship had deep roots in Danish history; in Viking times when polygamy was common these secondary wives were known as *slegfred*, an old Jutish title they could claim so long as they had spent three winters as a man's wife, holding the keys to his house and living there openly, sharing his bed and his table.

Choosing Kirsten was another definite step outside the boundaries of his class, one that would allow him far greater freedom to concentrate on his work. Morganatic marriages were not uncommon in this period, but they were usually second marriages, made for love when a noble first wife had died having fulfilled the needs of succession by providing legitimate heirs. Coincidentally, Frederick's son and heir Christian IV, who was instrumental in Tycho's exile from Denmark, married his very own Kirsten after his first wife died. Tycho, in love and careless of convention, was confident that he could overcome any difficulties and ignored his family's concerns. The problems came much later, as he struggled to guarantee the future of his projects on Hven by enabling his children to inherit Uraniborg and continue his work.

Portraits of Tycho show a proud, powerful man with typically sandy Scandinavian colouring – intensely blue eyes, clear golden skin and impressive blond moustaches that sweep down his chin at an angle. The Order of the Elephant, the highest honour in Denmark, hangs on a heavy chain around his neck. If you look closely, you may notice something strange about his nose – as an impetuous teenager, Tycho got into an argument with his cousin. During the duel that followed, it was sliced off and he wore a prosthetic one, made of metal, for the rest of his life. He kept it in place with salve which he kept in a little pot in his pocket and applied regularly. Several of Tycho's relations lost their lives this way; he was extremely lucky his wound did not get infected, and he spent several months recuperating. How did this brush with death affect him? On a practical level, he became very interested in medicine, particularly the ideas of Paracelsus and medical alchemy. If the rumour that he had foreseen a disaster that day in his astrological chart is true, it can only have strengthened his belief in the fundamental power and importance of the stars, while the brush with his own mortality and long convalescence may have increased his determination to devote his life to what he really wanted to do: astronomy.

Tycho's lifelong tendency to flout convention began at the tender age of two, when he was apparently kidnapped by his uncle, Jørgen Brahe. The circumstances aren't clear, and it appears that his parents, Otto Brahe and Beate Bille, had not agreed to give up their eldest son in advance. One explanation is that Inger Oxe, Jørgen's wife, couldn't have children, but she was only twenty years old so it's difficult to imagine this being unequivocal. Perhaps more likely, though jarring to our modern sensibilities, is the possibility that this was by design. Back then, young aristocrats were often brought up in other noble households, although it was usually something that was arranged and agreed by both parties in advance.

So Tycho did not grow up in the family home at Knutstorp Castle with his numerous younger siblings (his mother gave birth to twelve children in as many years – eight survived to adulthood), but instead was doted on as a precious only child. Jørgen and Inger lavished every kind of attention and opportunity on him; his childhood was a very different experience to that of his siblings as part of a large, boisterous family with a mother who was constantly weighed down by pregnancy. Despite this, Beate was a powerful figure in her eldest son's life, an exceptionally strong character who lived to the age of seventy-eight, no small achievement in those days, especially with all those pregnancies. The Billes were another prominent Danish family, whose love of learning was passed down to Tycho by his mother and her brother Steen.

Tycho's foster mother, Inger Oxe, also encouraged him in every way possible. Her brother Peder was a leading councillor to Frederick II, who sent him into exile when Tycho was about eleven. In 1566, he was dramatically recalled to restore Denmark's fortunes as Lord Treasurer. Oxe was the leading patron of learning and he encouraged everyone around him, including King Frederick, to follow suit. Urbane and cosmopolitan, he brought the sophistication of France and Italy to Denmark and showcased it at his beautiful Renaissance palace at Gisselfeld, surrounded by gardens, carp pools and exotic plants. Visitors enjoyed performances by the best musicians and exquisite delicacies like pheasant and oysters in rooms sumptuously decorated with paintings and tapestries.

Frederick II became king of Denmark and Norway in 1559, at a time when secular learning was just starting to flourish. The Reformation and the adoption of Protestantism in the 1530s had opened the country up to new ideas and, in particular, Philip Melanchthon's educational reforms. Frederick's parents engaged a professor of rhetoric from the university to tutor him and although he was clearly a bright child, he struggled to read and write, which they put down to laziness and stupidity rather than what we now understand as dyslexia. This caused people to write him off as illiterate and dim, and posterity to focus on his love of feasting, drinking and hunting.

Frederick was, in fact, an enlightened patron of culture, fully aware of the importance of the arts, crafts and sciences in making a country a respected player on the global stage. He put an emphasis on the quality of the design and production of every single item made in his country – something the Danes remain famous for today. He reformed the university in Copenhagen, providing generous funding to pay for teaching salaries and scholarships for poor students. He surrounded himself with learned men who shared his interests in alchemy, Paracelsian medicine and astrology. He brought Renaissance ideas to Denmark through the many foreign craftsmen he employed to build magnificent palaces for him, and the nobility followed suit.

Frederick's role as a patron reached its apogee in his relationship with Tycho Brahe, whom he liberally and enthusiastically supported throughout his reign. Tycho was unusual among scholars in so far as he was wealthy in his own right, but he could never have afforded to run a world-class research institute without royal support. Frederick granted him lucrative fiefs – lands, church livings and other income streams which brought in a constant supply of money. These were guaranteed while Frederick lived, but, as they were in the gift of the Danish crown and could be taken away as easily as they had been awarded, problems arose when Frederick died, and his son Christian became king. Tycho channelled this money downwards to the many people he employed in turn, making him an important scientific patron in his own right.

Like the Billes, the Oxe family was renowned for its passion for education – they were keen promoters of the new learning that had spread to Denmark from Italy and Germany. In contrast, Tycho's paternal family, the Brahes, were more traditional, boisterous rather than bookish, trained to govern and fight, to feast and hunt. They were country folk who spent much of their lives travelling between their fiefs, riding into battle when the need arose, managing their estates and attending the court. Between them, the ideals of these three families shaped Tycho's education and early life, the priorities of the Brahe clan sometimes clashing with those of the Billes and Oxes.

Tycho went to grammar school, where he quickly learned to read and write Latin. At twelve, he was ready to move on to Copenhagen University where he spent three years studying the traditional quadrivium syllabus. Astronomy was one of the four mathematical subjects, and Tycho's tutor Professor Scavenius taught a course based on Sacrobosco's classic text, *On the Spheres*. This was probably Tycho's first encounter with serious astronomical study, and while we don't know how quickly he became fascinated by it, the following year he did purchase two, more advanced, texts: *Cosmographia* by Peter Apian, which explained how to measure angles and make observations, and Regiomontanus' *On Triangles*, which demonstrated how to calculate horoscopes. The seed had been sown.

Tycho's next problem was persuading all four of his parents to allow him to continue studying astronomy, and for a while he kept his passion secret. By now he was a headstrong fifteen-year-old who had spent three years at the University of Copenhagen; it was time to move on. The traditional path for young noblemen, the one followed by his brothers, was to spend time as courtiers or knights at foreign courts, before returning home to learn the rules of governance in the service of the Danish crown. Tycho's education was different, thanks to the influence of the Oxe and Bille families. Peder Oxe had spent several years studying at a succession of European universities; Inger pushed for Tycho to follow in his footsteps. This decision would change the course of his life and enabled all of his later achievements, although astronomy could not have been further from the minds of his four parents when it came to Tycho's career. They imagined he was embarking on a degree in law that would prepare him to help govern Denmark. They chose a companion, Anders Sørensen Vedel, himself only twenty, and the two set off for Germany together on 14 February 1562. They travelled south on the well-trodden trade route that linked Scandinavia with Italy and southern Europe. On the way they passed through Lubeck and Wittenberg, reaching Leipzig in Saxony five weeks later – a long journey on horseback that must have been thrilling for Tycho, his first trip abroad.

Leipzig was one of the most important centres of trade in Europe, on the busy route that ran north from Italy over the Alps and on towards Scandinavia. In the twelfth century, fairs were established there by the fabulously named Otto the Rich; two hundred years later, the city was at the centre of an established international trade network, linked by road and the White Elster river and its two tributaries that flow into each other nearby. From here, boats could travel all the way to the North Sea at the port of Cuxhaven in northwestern Germany, having floated for several hundred miles on the rivers Saale and Elbe. When Tycho arrived, the central market square, overlooked by a grand newly built *Rathaus*, hosted the famous *Leipziger Messe* (trade fair) several times a year. Benefitting from special privileges and rules, these fairs were major events that attracted merchants from all over the world to sell their goods: woollen cloth from England and the Low Countries, food from across Germany, wine from the Mediterranean and exotic spices and luxuries from the east, painstakingly transported over the mountains from Venice, 600 miles to the south. Russian fur traders brought their wares through Poland, while Leipzig also lay on the major east/west trading axis, an important stopping point between Britain, France and Holland, and the eastern steppe. As always, trade in material goods was accompanied by the exchange of ideas and information. The University of Leipzig was founded in 1409, one of the first in Germany, and the printing press was well established there by 1480, making it a centre of knowledge production. Martin Luther visited the city several times and took part in the Leipzig debate, helping to inspire the widespread adoption of Protestantism amongst the local community.

Leipzig would have been an exciting place for a young scholar, especially someone who had grown up in the relative backwater of Denmark. For sixteen-year-old Tycho, it must have been an exhilarating experience, arriving in this bustling city, seeing all the shops and visiting the markets, being right in the centre of Europe, geographically and intellectually. Tycho's world began expanding in dramatic ways as he experienced new things, ideas and people.

In this sense, Tycho was like any young student arriving to begin their studies at the university, but of course he was different – there can't have been many others in his college whose parents were personal friends of the Electress of Saxony. No doubt Anne, as she was called, made Tycho welcome at court and kept a protective eye on his progress while he was in Leipzig. He certainly took advantage of the opportunities this entailed, making contacts and beginning to establish the network of like-minded intellectuals that would play such an important role in his professional life in the future. People like this were unusual in the sixteenth century, which must have made them all the keener to stick together and help each other out. The Saxon court astrologer, Valentine Thau, became a close friend; it's not difficult to imagine Tycho bombarding him with questions at grand dinners when he should have been making small talk with noble Saxon ladies.

Tycho revelled in the new world he found, making the most of everything it had to offer, including the shops. Like any young nobleman he had plenty of money to spend, but instead of heading to the tailor for elaborate clothes, he made a beeline for the booksellers and instrument workshops. One of the first things he purchased was a tiny celestial globe, small enough to fit into his fist, handy for slipping into his pocket and concealing from his tutor, Vedel, who forbade him to waste his time on astronomy. Tycho also bought Dürer's map of the constellations in the northern hemisphere, which he studied and learned by heart, using it to make rudimentary observations by using a piece of tightened string and checking the results against the information in *ephemerides*, annual diaries that gave the predicted positions of the planets, the moon and the sun, along with celestial events like eclipses. On 18 August 1563, aged seventeen, he began recording his observations in a special notebook. It was the first of many such notebooks he would fill with sketches, measurements and comments in his lifetime, the everyday vessels that carried his vast contribution to astronomy.

Vedel struggled valiantly to keep his young charge focused on his university course. Classical Latin and law just could not compete with the glittering universe above Tycho's head. He later admitted that he bought 'astronomical books secretly, and read them in secret', and spent his nights staring up at the skies, taking whatever measurements he could and writing down everything he saw in his notebook.[1] This was the beginning of Tycho's life as a scientific practitioner, the moment when he advanced beyond simply learning about astronomy and became an active participant in its development. For the time being, his stargazer's palace was the small chamber he had rented in the house of one of his professors; so long as he had a view of the night sky, he was happy. He must have been exhausted when it came to getting up for lectures in the morning, stars spinning around his head.

In the absence of any formal teaching in astronomy, men like Valentine Thau played a vital role in Tycho's intellectual development, providing inspiration and information, fuel for his passion which lay outside his formal study of law – the subject Leipzig University was famous for; indeed, the reason Peder Oxe had sent him there. Tycho later explained that 'in fact I never had the benefit of a teacher in mathematics [astronomy], otherwise I might have made quicker and better progress'.[2] While this obviously frustrated him, it did have various positive effects. Because he was effectively teaching himself, he was free to think and pursue the subject in any way he wished.

Traditional astronomy professors discouraged looking at the stars and relied on information provided in *ephemerides*, but Tycho, working alone in his room, followed his own path. He began by using an edition of *ephemerides* calculated by an Italian called Giovanni Battista Carelli in 1561 using the Alfonsine Tables made in the fourteenth century, based on Ptolemy's model with the earth at the centre of the universe. Tycho then demonstrated his rigorous working methods by purchasing an alternative set of *ephemerides* by Johannes Stadius to use as a comparison. These used the newer Prutenic Tables, which drew from Copernicus' improved calculations that took the tilt of the earth's axis into account and were based on his heliocentric model of the universe. Tycho was shocked by the margin of error between the predictions and what he actually observed in the night sky. On 23 August 1563 he watched the conjunction of Saturn and Jupiter, the two planets passing so close to one another they seemed to be one astonishing source of light. Carelli's prediction was that this would not occur until 17 September, while Stadius had it marked for 24 August. Tycho wrote that from then on, he no longer trusted the *ephemerides*.[3] This was the moment when he realised that accurate observation was the key to successful astronomy, and he needed to spend more time looking up at the heavens, not down at books. It went without saying that accuracy relied on high-quality instruments – pieces of string would no longer do.

At this point, Tycho's friend and fellow student Bartholomew Schultz (Latinised to Scultetus) came into the picture. He was engaged in a pioneering project to survey and map Saxony; he introduced Tycho to the related disciplines of geography, cartography, navigation and, most important of all, the design and production of instruments with which to carry them out. Instruments would become the vital force in Tycho's scientific life. If his notebooks were the vessels of his scientific achievement, his instruments were the apparatus on which it hung. Scultetus showed him how to use a cross-staff or radius to measure the angles of separation between individual stars and helped him to make one of his own, 'according to the direction of Gemma Frisius'.[4] Tycho made his first observations with it on 1 May 1564, measuring the angles between four of the planets. His excitement was palpable: 'When I had got this radius, I eagerly set about making stellar observations whenever I had the benefit of a clear sky, I often stayed awake the whole night through, watching them through the window of one floor, while my [tutor] slept and knew nothing about it.'[5]

After a brief time back in Denmark, at that time mired in war, Tycho went to the great centre of northern humanism, Wittenberg, where he studied under Caspar Peucer. This fruitful period was cut short when an outbreak of plague forced him back onto the road, this time to Rostock and the fateful duel. As soon as he had recovered, he headed south, visiting first Wittenberg and then on to Basel where he matriculated at the university. In the late 1560s he spent fourteen months in Augsburg, where he lodged with a goldsmith, who initiated him into the mysteries of Paracelsian chemistry, a passion that would match his love of astronomy. Experimenting in the workshop opened not only the world of alchemy, but also of fashioning things out of metal.

Tycho settled happily into a circle of older, erudite men who, with their eclectic interests, introduced him to a new way of living an intellectual lifestyle. Paul Hainzel and his brother Johannes were the prime movers in this group and their collections of instruments and automata opened Tycho's eyes to the potential of technology. He was frustrated by the errors his cross-staff produced. Some were systematic and he wrote up tables of corrections, but others were random and therefore impossible to predict and rectify. He had a large pair of compasses made, big enough to take detailed measurements. The problem was that without anything to mount them on to keep them still and steady, errors still occurred. He realised that he needed a much larger instrument, fixed onto a stable base. One day he was in the street discussing this problem when he bumped into Paul Hainzel, and Hainzel was so taken by Tycho's passion for this idea he agreed to provide the money to pay for it and suggested building it on his family estate outside the city. For the first time in his life, Tycho was able to see his dreams becoming reality.

The resulting quadrant was vast, with a radius of five and a half metres set into a solid oak base so heavy it took forty men to move it. While it did produce readings of improved accuracy, its size and weight made it far from easy to use. Yet the *Quadrans maximus* was an important step on Tycho's path as an innovator in instrument design and production – something that really flourished once he was master of his own observatories on Hven. Hainzel's support made him aware of the possibilities that generous patronage could provide, making him more open to Frederick II's overtures on his return to Denmark. While in Augsburg he also commissioned a vast brass celestial globe and met the French humanist Peter Ramus – like Dee in Louvain, the city had a transformative effect on him. Meanwhile the journey was turning into an astronomical grand tour. Tycho had visited Cyprianus Leovitius on his way to Augsburg (and been astounded to discover he never made observations), and on his way north to Nuremberg he stayed with Philip Apian (son of Peter) in Ingolstadt. In Nuremberg, he purchased a cross-staff and a set of rings made by Walter Arsenius in the Louvain workshop formerly run by Gemma Frisius. While in town he would have seen Dürer's house and been able to look up at the balcony on which Bernhard Walther had made observations – the city was full of inspiration for a young man with astronomical interests.

By December 1570 he was back home. In May he took his place beside his brothers at their father's deathbed, and the next few months were spent sorting out the vast Brahe inheritance, which was shared between his sisters, his widow and their children. Now enjoying a period of peace, the country flourished under the careful stewardship of Peder Oxe. Money flowed into the country, as did Renaissance ideas and luxuries. Tycho began spending most of his time at Herrevad Abbey with his uncle, Steen Bille, one of the few in his family who appreciated and encouraged his scientific interests. Steen ran the former monastery which still had its distilling house, library and apothecary, while the old monks remained to teach young Lutheran boys at the school – a far cry from the destruction and brutality of the dissolution of the monasteries in Britain and elsewhere. He oversaw the estate workshops and mills along the Rönne river that produced bricks by the thousand and wheelbarrows by the hundred, and he was also in charge of the ironworks that supplied the weapon factories over the Sound in Copenhagen.

Steen added to these by inviting a German called Rotther to found the first Danish paper mill nearby, and persuaded Frederick II to send his court glassmaker, a Venetian, to set up a glass works at Herrevad. It is no wonder Tycho moved in. Here he could visit the forges and watch iron and steel being made, glass being blown and paper manufactured before his very eyes. There were brass workers, blacksmiths, cabinetmakers, all ready to embark upon the production of whichever new-fangled instrument he wanted, his own Augsburg in miniature. The glassworks supplied him with vessels and flasks for the laboratory Steen had allowed him to set up in one of the abbey outbuildings. Astronomy took a back seat – there are no extant observations and, for a while, he seems to have been totally focused on alchemy.

The night sky soon came calling, however. On the evening of 11 November 1572, Tycho was walking back from the laboratory for supper when something caught his eye – an astonishingly bright new star. It was the supernova that today bears his name, 'truly the greatest miracle in the whole of Nature since the beginning of the world'.[6] The alchemical furnaces were dampened as Tycho's eyes were resolutely trained upwards, on the constellation of Cassiopeia. A few years earlier he had invented a new instrument called a sextant; he had one made and mounted onto one of the windows of the abbey, and used it to make observations, night after night, comparing and checking his results and measuring the angles between the supernova and other celestial bodies. He was trying to ascertain whether there was parallax, apparent movement in the star during the hours of the night. If there was, this would suggest it was closer to the earth (the closer the object, the larger the parallax). It remained visible for eighteen months, but began to dim soon after it appeared, the colour gradually shifting from white, through yellow to red.

Tycho collated and processed all of his data using Regio-montanus' theory of triangles to calculate the longitude and latitude, but found no discernible parallax (however, others did), and this forced him to conclude that the new star was far above the moon, putting Aristotle's theory of a perfect and unchanging heavenly realm in doubt. Tycho wrote up all this information in a book called *De Nova Stella*, which he showed to his close friends. He initially refused to publish the book because it would be considered ungentlemanly, and for a while toyed with the idea of publishing anonymously. He was young and inexperienced, nervous about his peers' reactions, especially since his calculations placed the star much further away from earth than many others thought. Eventually, Peder Oxe weighed in and insisted he publish it. The compromise was a limited print run of which very few copies survive, only disseminated to a small, elite group of astronomers.

De Nova Stella began with letters between Tycho and his friend Johannes Pratensis (a professor at Copenhagen University) explaining these concerns, along with a detailed description of the star, its position and attributes. The next section focuses on its astrological significance; given the widespread instability caused by religious strife (1572 was the year of the St Bartholomew's Day Massacre in Paris), it is not surprising he concluded that the star presaged tumult and misfortune. Comets were also generally held to be harbingers of misfortune and conflict.

Tycho's anxieties about making his observations public were understandable. There were huge variations in observations of the supernova. Tadeas Hajek's measurements were out by as many as 16°, Thomas Digges' by up to 4°, Cornelius Gemma by around 1°; Peucer found a parallax of 19°, while Tycho and Wilhelm IV found none, convincing them that the star was much further from earth.

Michael Maestlin, who taught Kepler and followed Copernicus, owned no instruments but found no parallax either, and was able to locate it quite accurately.

Lively debates followed as scholars compared data, methods and instruments, arguing over where the supernova was, what it was made of and what it meant. Some, including Wilhelm IV, suggested that because the star of Bethlehem had accompanied the birth of Christ, this new star predicted the last judgement and the end of the world. Naturally, Catholics claimed it signalled victory for the true faith, and everyone consulted their history books to find precedents to back up their theories. It was a transformative moment in the history of science, the first time that a diverse group of scholars in such a wide range of locations had discussed their findings on a single subject. Astronomy was centre stage, no longer merely the 'handmaiden' of astrology or an impenetrable sequence of mathematical models – now it was focused on the fundamental questions about the universe and how it works. For Tycho, this was his formal entry into the intellectual world. No longer merely a student, he was now a fully-fledged member of an elite group of people interested in the natural world.

By this time Tycho was living with Kirsten, and they were the proud parents of a baby daughter. The other significant woman in his life was his sister Sophie, the youngest of the ten Brahe siblings. Still a teenager, she lived with their mother not far from Herrevad and in 1573 she helped him to observe an eclipse of the moon using the gilt brass quadrant he had recently commissioned from the royal clockmaker Stephen Brenner, who was originally from Nuremberg. If studying astronomy was unusual for a Danish nobleman, it was unheard of for a noblewoman; Sophie was not just swimming against the tide, she was in a different ocean altogether. Tycho's father had been reluctant to allow his sons to learn Latin, so it's not surprising that his youngest daughter knew none. She did not let this stop her, and paid for books on astronomy to be translated into Danish or German so she could read them. She was also interested in judicial astrology, although Tycho warned her the female mind was not capable of comprehending such a complex subject. Sophie took no notice and taught herself how to cast horoscopes. Tycho's astonishment at his sister's intelligence is almost touching; he clearly adored her and she played a vital role in life on Hven in later years when she stood in as hostess – Kirsten's status disbarred her from taking on this role when noble guests were staying.

The same year, Tycho articulated his views on astrology in a lecture at the University of Copenhagen, highlighting that the influence of the stars is through God and cannot be denied, while trying to address the criticisms of preceding decades. He described how the organs are influenced by the planets: the liver by Jupiter, the kidneys by Venus, the lungs by Mercury, and that the sign you are born under determines your fate and characteristics. Venus denoted love, music and pleasure, Mercury business, the moon travel. At the same time, he allowed for the possibility that people could conquer these influences if they were strong enough.

In 1575, Frederick sent Tycho on a mission to Italy to find craftsmen who would come back to Denmark and help build his new palace of Kronborg. Unbeknownst to Frederick, Tycho had his own motive for wanting to travel – he was on the lookout for somewhere to settle down, away from Denmark and 'this way of life, these customs, this daily chatter'.[7] As we know, he visited Kassel but was not able to stay as long as planned. Leaving Wilhelm's court he headed south, towards the Alps, which he needed to cross to get to Italy. He stopped in Frankfurt, and by lucky chance the trade fair was on, so he bought several books and pamphlets on the new star. Of all the places he passed through, Basel, a picture-perfect town on the Rhine in Switzerland, framed by snow-capped mountains, was the place where he felt most at ease. Blessed by political neutrality and a reputation for promoting peace, the city was a haven for intellectual thinkers and one of the crucibles of humanism. Erasmus had found shelter there a few decades earlier, giving impetus to the burgeoning printing industry alongside his friend Johann Froben whose publishing house quickly became a byword for quality. In 1543, Andreas Vesalius had printed his revolutionary book on human anatomy in Basel, cementing the city's status as a centre of innovation and excellence, so it was hardly surprising Tycho felt at home there. This was also his second visit, so there were acquaintances waiting to welcome him.

He did not spend long in Italy, just enough time to find some artists in Venice and offer them Frederick's patronage if they came to Denmark. Then he was off again, back over the mountains to Augsburg where he checked on the progress of the globe he had commissioned and visited the Hainzel brothers. From here he joined the crowds travelling to Ratisbon to see Rudolf II being crowned King of the Romans. He hoped to cross paths with Wilhelm IV again but was disappointed. However, he did meet the imperial court physician, Tadeas Hajek, a central figure in the European astronomical community who Tycho corresponded with for decades. They exchanged ideas – Hajek gave him a manuscript of an unpublished work by Copernicus describing his new world system, and in return Tycho presented him with a copy of *De Nova Stella*. They discussed the phenomenon and reactions to it, and Hajek also passed on a letter from an astronomer in Spain, forging another link in the Europe-wide community of scholars.

Tycho's next stop was Saalfeld to see yet another astronomer, Erasmus Reinhold the younger, whose father had compiled the Prutenic Tables, the most up-to-date astronomical data available. Tycho was shown around the library and Reinhold senior's manuscripts, giving him insights into how the tables had been calculated. From here he passed through Wittenberg, in crisis after the suspected Calvinist plot, where he saw some old friends and new instruments, and discussed the supernova. By the end of the year, he was back at home with Kirsten.

History doesn't relate what Kirsten thought of the proposed move to Basel. Frustratingly, inevitably, she is a silent figure who hovers just behind Tycho, running his household, having his children, being a loving wife. Would she have been anxious about the prospect of leaving for another country with her baby daughter in her arms? Of moving to a place where she couldn't speak the language, had no friends or family? As it turned out, she needn't have worried, at least not for another two decades. Wilhelm IV praised the young astronomer and strongly advised Frederick to make sure he settled in Denmark – Tycho was summoned to a royal hunting lodge near Copenhagen where Frederick made him an offer he couldn't refuse: 'I saw the little island of Hven, lying in the Sound in the direction of Landskrona…if you want to live on the island, I would gladly grant it to you. There, you can live peacefully and carry out without disturbance the studies that interest you.'[8]

Thanks to Wilhelm, Frederick could see what an asset Tycho would be to Denmark, and could envisage what a first-class research institute would do for the country's reputation and international standing. Tycho set to work at once, intent on recreating the centre at Herrevad, but on a much grander scale. Unfortunately for them, the inhabitants of Hven were central to that plan. Until now, the island had been part of the general property of the Danish crown; in the absence of a lord in situ to run things, the islanders had largely been left to their own devices and were used to organising their lives as they wished. Now, they had to work for Tycho for free two days a week, and they were far from thrilled to be caught up in his grandiose and often crazy schemes. Danish society was not so rigidly structured as in other European countries, just as the Danish monarchy, at this point, was not absolutist but ruled in conjunction with a council of nobles. As free-holder farmers, the inhabitants of Hven paid tithes to the Church, made obeisance to the king and ran things through well-established social structures. They had no desire to give up their freedoms to a capricious, arrogant young landlord who appeared to think he owned them and their island outright.

The proclamation from Frederick II was read out to the islanders on Tycho's second visit, announcing that Hven had been awarded to him for his lifetime. Assembled in Tuna, the tiny village that was the island's only settlement, the locals noticed with disapproval that their new lord did not remove his hat as a mark of respect. Their lives were about to change, but they would not give up their freedoms without a fight. The clause that would echo around the island throughout the following decades read, 'But he shall observe the law and due right towards the peasants living there, and do them no injustice against the law, nor burden them with any new dues or other uncustomary innovations.'[9] The irony of the last two words is impossible to ignore – Tycho's whole project on Hven centred on designing, developing and building 'uncustomary innovations' in science and he intended to run the island estate as an organised, centralised entity, a system known as *Gutswirtschaft*, a system that was on the rise across the country. This involved taking ownership of the peasants' land in order to increase agricultural production and sell the surplus to foreign markets; it was underpinned by the transfer of power from the farmers to a bailiff, crucially requiring them not only to pay the usual dues, but to work without pay for the landlord a couple of days every week as well.

Unpopular as it was, *Gutswirtschaft* contributed significantly to Denmark's development into an agricultural superpower, one of the biggest exporters of meat and dairy products in the world – Danish farmers kept food on many British tables during the Second World War and its aftermath. The Hvenians were devastated to have to surrender the lands they had passed down for generations, but as their new landlord arrogantly pointed out, they had 'no documents, privileges or freedoms' to demonstrate their ownership.

As there was no great house on Hven for Tycho to move into, Frederick II generously gave him the enormous sum of 500 dalers to put towards building one. Tycho strode to the high ground in the centre of the island from where there is a commanding 360-degree view of the mainland on each side and the sea to the north and the south. He announced that this would be the site of his new demesne, completely disregarding the fact that the land was currently held in common for grazing by the local community. Work started almost immediately; the French ambassador, Charles Dançay, laid the cornerstone early in the morning of 8 August 1576, and libations of wine were made and prayers for success sent heavenwards. Tycho employed a Dutch master builder, an Italian architect and a German painter to oversee the work, the beginning of a fluid, multicultural community that grew over time.

Uraniborg, the 'Castle of Urania', named for the muse of astronomy, was a stunning Renaissance palace, a luxurious family home complete with studies, observatories and laboratories. The design was perfectly symmetrical, set within a diamond-shaped enclosure with gates on four sides and avenues leading to each of the four facades of the building. A deep band of trees was planted inside the perimeter walls and the spaces between these and the circular courtyard that surrounded the palace were divided into intricately patterned flower beds. A vision of architectural beauty and scientific utility that was inspired by Tycho's visit to the Veneto in 1575, there are striking similarities in the ground plans of Uraniborg and Andrea Palladio's newly built Villa Rotonda.

Tycho based many practical aspects of Uraniborg on his uncle's research centre at Herrevad Abbey, but what were his other inspirations? Where did he get the idea for so much splendour, symmetry and symbolism? There was an abundance of grand estates with great houses, but few that also functioned as research institutes. Tycho had travelled extensively in Germany, initially as a student and later as a scholar; in both instances he spent time at courts of local rulers as well as in the studies of the intellectual communities and university halls. His visit to Kassel was certainly influential; it showed him what was possible with money and imagination, but also the importance of being able to fully focus on astronomy, as Wilhelm was unable to do, and to design an observatory space where large instruments could be used and stored, without the need to move them. There were a growing number of Renaissance-style houses in northern Europe, as the influence of Italy spread. Peder Oxe's elegant house at Gisselfeld and Hainzel's estate outside Augsburg certainly played into Tycho's plans for Hven.

Frederick and Tycho got on well; they had many things in common. They were both happy to do away with the formalities of court life and treat people as their equals, a Danish quality that is still an integral part of the national character. Pupils in Danish schools call the teachers by their first name and woe betide anyone who thinks they are above anyone else and breaks the law of Jante.* Both men were interested in people for their talents rather than their social position, and endowed with so much confidence that they could break free from the constraints that ran through every stratum of early modern society. At feasts, Frederick would bellow, 'the king is not at home', a signal that everyone could relax and get on with the serious business of partying. When it came to employing people to work at Uraniborg, anyone with the ability and the passion could come and join Tycho's 'familia'; the pool of talent on offer was as large as possible. In an era when the intellectual world was so tiny in terms of numbers, this was essential. Hven had the reputation of being welcoming, and young men came from far and wide to work alongside the legendary astronomer on his extraordinary island.

Tycho spent much of his time on Hven during the construction work, watching his vision slowly solidifying into bricks and stone. He still found the time for relaxation and one afternoon in November 1577, he went fishing. As the sun began to slip down behind the Danish mainland to the west, he noticed strange red rays stretching across the horizon. For the first time in his life, he was looking at a comet. He rushed back to fetch his cross-staff so that he could measure it; the next day he wrote to court explaining that he would not be able to leave Hven for some time. For the next two months he was completely focused on following the comet's progress, observing it whenever the weather allowed using the steel sextant he had invented and his azimuth quadrant. He

---

*   '*Janteløven*' or 'The Ten Rules of Jante' were formulated in the 1930s to encapsulate the conformity and egalitarianism of Nordic cultures. Rule 1 is 'Don't think you are anything special.'

calculated its parallax and then sat down to write a secret report, in German, for King Frederick and Queen Sophie. He described it for them, claiming that the colour of the tail came from the sun's rays and that it was far above the moon, close to the orbit of Venus. It was 'tremendously huge and had a malicious, saturnine appearance' and would certainly cause havoc on earth, especially once the great conjunction of 1583 occurred, at which point 'it is expected that the eternal Sabbath of all creatures is at hand'.[10] No wonder he marked this treatise for the eyes of the royal couple alone; the outlook was not rosy.

Tycho and Frederick both believed passionately in astrology and the heavens' influence on earth. One of the first things the king did on the birth of his son and heir Christian was ask Tycho to prepare the baby's horoscope. (Casting nativities was a major source of income for professional astronomers; much of the impetus for improving observational accuracy was geared towards better readings of the heavens for astrological purposes.) Tycho claimed to have calculated a horoscope for Casper Peucer in 1564 that predicted he would spend several years either in prison or in exile and would regain his freedom at the age of sixty – which of course turned out to be correct. Instances like this naturally increased his confidence in the power of astrology, but he was always keen to practise it with rigour and precision. He made the same observations with different instruments and calculated the average, a much more scientific approach than was usual at that time.

The last months of 1577 were feverish for the whole astronomical community. Everyone, even little Johannes Kepler, just six years old and shivering in the night air, spent that winter gazing heavenwards. In the following decade, more than one hundred books were written on the comet, discussing where it was, what it meant, how it came to be and what it was made of, and the scholarly community were united in their confusion like never before. Almost every text contained predictions about what the comet meant for life on earth. Only four were anti-astrological, presenting a range of often conflicting viewpoints ranging from a pro-Aristotelian who believed that agriculture was the only area where astrology could be applied to the more pragmatic theory that comets were just another form of natural phenomena with no effect on earth.

Andreas Dudith voiced the latter opinion in his *Little Commentary on the Meaning of Comets*, a slim volume which took aim at the superstitions of mainstream astrology. This skilled physician, born in Hungary of Italo-Croatian parents, and educated in France and Italy, was consequently in touch with people all over the continent. As far as he was concerned, plagues had medical, not astrological, causes, and kings died, wars raged and earthquakes struck regardless of whether a comet had appeared; his wide critical reading of writers from Ptolemy to his contemporary Girolamo Cardano made him a formidable intellectual opponent.

Many of Dudith's close friends felt wounded by his book, and none more so than Hajek, who responded with accusations of impiety, something Dudith had anticipated in the prologue to the treatise: 'Moreover if one is guided by solid rational arguments and by...experience, and comes to the conclusion to abandon the common and popular opinion, superstitious and exaggeratedly fearful people will accuse him of the serious and horrible sin of impiety.'[11] Perhaps inevitably, the argument twisted and warped into the role God played; by attacking the fundamental basis of astrology, Dudith revealed deep chasms in sixteenth-century belief. He was certainly not the first to do this – there had been attacks on astrology for decades – but the popularity of his book well into the seventeenth century shows that these ideas were gradually gaining purchase, as people were becoming increasingly willing to question the fundamental structures of their cultural beliefs.

The reaction to the comet of 1577 was immediate, but the intellectual change it provoked lasted several decades. Like the new star of 1572, it drew attention to the night sky, and people began to contemplate the mysteries of the universe with new energy and in new ways. Astronomers compared their measurements and discussed the discrepancies; they argued over what the comet portended for life on earth and 'all the differences of opinions that divide learned men in several ways'.[12] These discussions and disagreements were extremely valuable – the first time in western Europe that scholars from such a wide range of countries conferred in this way. Hven, along with Kassel, London and Prague, were major hubs on this new map of knowledge.

By November 1580, the palace of Uraniborg was complete enough for Tycho, his family and assistants to move in. The first building in Scandinavia in the Gothic Renaissance style – a heady combination of Italian and Dutch architecture, it was built of brick with sandstone flourishes, gables and domed turrets, curved glass glinting in the windows. It was exquisite, small yet perfectly proportioned and elegantly decorated, the ultimate stargazer's palace. As practical as it was beautiful, every room had been specially designed to fulfil a purpose. Tycho's library or 'museum' was on the ground floor of the south tower, the great celestial globe from Augsburg glinting in the pale sunlight alongside his collection of automata. The kitchen was in the other tower, while the central square in between was divided into four rooms, the Brahes' great four-poster bed standing in one, with a huge table made of oak.

Upstairs, the chambers were each decorated a different colour – red, blue, yellow and green. This floor was topped with a domed octagonal pavilion. The green room's ceiling was painted with plants and flowers, with a beautiful view over the Sound to the west. In the attic above were smaller rooms for assistants and students. The cellar contained Tycho's alchemical laboratory complete with furnaces, along with storage rooms for food, wine, salt and fuel. A complex system of lead pipes designed by a crafts-man from Nuremberg carried running water to both the main floors and fed the gargoyle fountain in the main hall. The main rooms were warmed by brightly coloured, tiled stoves (all of the firewood for which had to be imported from the mainland, as did everything else, from food to musicians). There were also studies and smaller workspaces for students, a second laboratory and a winter dining room. Inscriptions, cornicing, portraits of famous astronomers and allegorical figures decorated the walls, creating an inspirational atmosphere.

Astronomical instruments weren't confined to a single room at Uraniborg – they were everywhere. Some were positioned in the upper rooms of the two towers, linked to smaller observation plat-forms by galleries and stairs. The roofs were constructed with triangular boards that could be removed to expose the sky, so the instruments could be pointed at any celestial body Tycho wished to measure. This allowed him to use huge immovable instruments, like his large sextant and the wall-mounted quadrant, which needed to be both protected from the elements and trained on the night sky. There are echoes of Wilhelm's balconies here, taken to another level at Uraniborg where the entire roof structure was effectively an observatory. The small portable instruments were kept close at hand, so they could be grabbed and taken out onto the galleries and platforms at any moment, but it was the giant instruments that would play a central role in Tycho's observational programme. He had concluded early in his career that they would enable him to make his observations far more accurate, and so, as soon as the building was complete, the installation of these instruments began. These included a large quadrant mounted on a ball and socket so that it could be swivelled to different positions, a huge steel azimuth quadrant for the southern tower room, various sextants and a vast 3.3-metre triquetrum in the north rotunda.

The famous mural quadrant, with a 194-centimetre radius, was installed on the ground floor in the room next door to the library, oriented due south. It soon proved its worth by producing data of unparalleled precision. These 'in situ' instruments were naturally supplemented by a huge array of smaller, handheld ones, cross-staffs, quadrants and astrolabes, although Tycho did not use the latter for making observations because he considered them inaccurate. All the instruments, except for those from Augsburg or Louvain, were made in the palace workshop on the edge of the garden, overseen by an exceptional instrument maker from Westphalia named Hans Crol. This gave Tycho unprecedented power over his instruments; they were designed, made and altered right there at Uraniborg under his watchful eye.

In the high summer of 1580, a few months before Tycho's familia moved into Uraniborg, Paul Wittich arrived on the shores of Hven, a letter of recommendation from Hajek in his pocket. It was not long before he and Tycho were deep in discussion, pouring over books in the Uraniborg library and Wittich's manuscript copy of Copernicus' *Commentariolus*, in the author's own hand. This was a gift from his uncle, a Polish scholar who had studied with Georg Rheticus, Copernicus' only pupil and the man who had persuaded him to publish his masterwork. This connection impressed Tycho, as did Wittich's mathematical abilities, but they disagreed on the location of the new star and the comet of 1577. Wittich, who did not often make observations, probably because of his poor eyesight, thought they were below the moon.

As luck would have it, a comet appeared on 10 October. Everyone rushed to observe and plot its journey across the heavens; even Wittich was pressed into service despite the fact that he was 'not at all adept in using these [instruments] and could not even take one observation either of a comet or a star'.[13] He made up for this once the data was in and the calculating began, showing the Hven astronomers the method of prosthaphæresis he had learned from Rheticus which vastly reduced the time and effort required. Tycho immediately set him to work writing a manual explaining trigonometry to keep at Uraniborg; perhaps he had heard about Wittich's propensity to wander, and wanted to make sure he did not lose this new technique. Sure enough, at the beginning of November, Wittich stopped writing the trigonometry manual and told his new colleagues that he had to return to Wrocław because his uncle was ill, promising he would return before long. Tycho, happy to have the prosthaphæresis method, was generous and gave him a copy of Apian's *Astronomicum Caesareum* worth forty grams of gold, no doubt with the expectation of loyalty and his presence again in the future.

Wittich never did return to Denmark. He spent the next couple of years working with Andreas Dudith and a fluid circle of scholars in Wrocław, including the Englishman Henry Savile and the Scot, Duncan Liddel. They discussed the competing merits of the Copernican and Ptolemaic systems, and compared their ideas with other astronomers via letter, notably Hajek in Prague. Notes and diagrams illustrating these discussions and showing alternative planetary systems have been found in various copies of Copernicus' *De Revolutionibus*. Initially historians assumed the annotations were added by Tycho himself, but recent investigations have shown that they are in fact the work of the mysterious Wittich, who owned four copies of the book, all of which contain notes. The situation took an unforeseen turn on 9 January 1586, however, when Wittich died suddenly. He had published nothing, leaving the field open to anyone who wanted to claim his work for themselves. An unseemly scramble ensued as his peers tried to get their hands on his library and papers, which had gone to his sister after his death. No one was more relentless in this quest than Tycho, who continued to pursue Wittich's books until his own death in 1601 and used his own publications to assert his singular authorship of the Tychonic system and other related ideas. In the process, he almost totally obscured Paul Wittich's contribution to sixteenth-century astronomy.

Hven was ideally placed to give Tycho privacy while still being easily accessible. Visitors came from far and wide. Mathematicians like Wittich and Rothmann from Germany and Poland, Tycho's old university friends, Vedel, Pratensis and others, often sailed over from Copenhagen, and assistants came from France, Norway, Scotland and beyond. Tycho's noble friends and relations came to admire his beautiful palace and feast on elaborate delicacies specially shipped over from the mainland. Queen Sophie was so impressed by her visit in the summer of 1586, she returned a month later with her parents. Foreign diplomats often dropped in on their way back southwards from Copenhagen, and in 1590 James VI of Scotland brought his new bride Anne, Frederick and Sophie's daughter, for the day before they set off to start their lives together in Britain.

The most frequent visitor, however, was Tycho's beloved sister Sophie. Her husband died in 1588, leaving her in charge of Eriksholm Castle and their little son, Tage. This left her free to pursue the intellectual studies she had been interested in since she was young. Now a widow in charge of her own destiny, she could spend time making medicines in her distilling house using the numerous plants she grew in her renowned garden at Eriksholm. In 1589 she helped oversee a new planting scheme at Uraniborg, adding medicinal herbs like angelica, wormwood, juniper and saffron crocuses to the box-edged beds around the palace.

Another reason for Sophie's many visits to Hven was the presence of Erik Lange, a nobleman who was so fanatical about alchemy he bankrupted himself. He and Sophie fell in love, and they eventually married, in penury, after a decade-long engagement. In September 1584, Lange spent two weeks on Hven with Tycho discussing the latest astronomical ideas. The only fly in the ointment was a young German scholar called Nicolaus Reimers Baer, who Lange brought with him as his mathematician. Baer, later known as Ursus ('Bear'), behaved suspiciously, snooping through Tycho's papers, messing around with instruments and losing his temper at the slightest provocation. Tycho wound him up by claiming, 'All those German fellows are half cracked,' an insult Ursus never forgot. Distrustful, Tycho asked one of his assistants to search Ursus' room for stolen information, and he found 'whole handfuls of tracings and writings.'[14] Ursus reacted violently when confronted and Lange dismissed him; the party left Hven under a cloud. The scene was set for the astronomical quarrel of the century.

In the light-filled summer nights when the sun had hardly set before it began to rise again, the celestial bodies were only visible for a few hours; the darkness of the winter months was better suited to star gazing. The downside of Hven's location was its exposure to the wild Nordic climate, and in particular, the wind that rips down the Øresund on a regular basis. Keeping the instruments aligned and steady on the viewing platforms was challenging in all but the calmest weather and bearing in mind that the night also had to be clear, this seriously limited the opportunities for observation. Tycho began to realise that his plan of collecting a large, consistent body of observational data on a rooftop observatory in the far north of Europe was unrealistic, so he considered other options. The solution he came up with in 1584 was to build, or rather excavate, another observatory arranged within three underground crypts (two more were added in 1588) around a central study, just next to Uraniborg. Beds were installed for the long nights waiting for the weather to change, and there were retractable domes above the instruments, which were now stable and relatively protected from the weather. In Stjerneborg (Star Castle), Tycho and his assistants could make observations more frequently, and with greater precision.

The same year, Tycho began building his own printing press on the island. Frustrated by the lack of scientific printers in Copenhagen and increasingly concerned about accuracy and plagiarism, he hired a printer and, following in Regiomontanus' footsteps, began publishing his own books. The next problem was the supply of paper, which had to be brought from places as far afield as Frankfurt, a complicated and expensive process. Tycho's response was typical – he would build a paper mill. With no river on Hven, an expert surveyor was brought from Kronborg to create a system of ponds connected by dams and sluices to create sufficient water pressure to power it, and this process took some time to complete. Once it was finally up and running, the long-suffering islanders were called upon to donate their bedsheets to make the paper, adding to their long list of grievances.

In 1588, Tycho wrote to his old mentor Caspar Peucer to explain the ambitious publishing project he was going to call *The*

Theatre of Astronomy. It would begin with an *Introduction* in three volumes, followed by seven books each focusing on a different aspect of the new astronomy: instruments, trigonometry, prosthapæresis, a new star catalogue and so on. Through most of 1587 the Uraniborg Press was busy printing the first book of the *Introduction*, Tycho's long-awaited work on the comet of 1577, based on his own meticulous observations and a comprehensive survey of other works on the topic. In the eighth chapter, he revealed his masterwork, the Tychonic system of the universe, which was a fiddly compromise between the Ptolemaic and Copernican cosmoses. A total of 1,500 books were printed, and copies sent to a select group of astronomers. A few were also sold at the Frankfurt Fair, but most were stored in readiness for the rest of the project to be printed. Meanwhile, Ursus had been in Kassel in early 1586 where he showed Wilhelm, Rothmann and Bürgi the new planetary system he had been working on. Wilhelm was so impressed, he asked Bürgi to make a model of it. In 1588 Tycho gave one of his scholars who was travelling to Germany a copy of the *Introduction* to deliver to the astronomers at Kassel. Not long after, word came back from Rothmann that Ursus had visited the year before and shown them a very similar system which he claimed as his own. Tycho was beside himself with rage; his worst fears had come true – Ursus, 'savage, inhuman, scurrilous, rotten and sycophantic', had stolen his masterpiece.[15] The ensuing row spanned several years and thousands of miles. Astronomers weighed in on every side and Tycho's mission to establish his authorship took on a new urgency.

Before long, a much worse storm began to brew. Tycho's days on Hven, safe in his scientific kingdom with his instruments, his assistants and his printing press, were numbered. We will meet up with him again in Prague, by which time he had managed to publish detailed records of his observatory, his instruments and the work he did on Hven, preserving and disseminating his legacy to the next generation of scholars. This was fortunate because the stargazer's palace he built did not last long before it was reduced to rubble, the stones taken by the islanders for their own houses. They were keen to forget their tyrannical landlord, but the books he published ensured that the name of Tycho Brahe, and that of Uraniborg, lived on.

6

PRAGUE

THE CITY AT THE CENTRE
OF THE WORLD

His majesty is only interested in wizards, alchemists, Kabbalists
and the like, sparing no expense to find all kinds of treasurers,
learn secrets and use scandalous ways of harming his enemies...
He also has a whole library of magic books.

The Archdukes of Austria on their ruler, the Holy Roman
Emperor Rudolf II, 1606

A s we draw near to the close of the sixteenth century, it
starts to feel like all roads are leading to Prague. Instrument
makers, astronomers, artists, botanists and alchemists
found themselves caught by the magnetic pull of Rudolf II's court,
its reputation for cultural freedom and enlightened patronage.
There was no better place to be in all of Europe, no greater patron
than His Imperial Majesty. As a result, modern historical studies
can appear overwhelming, physically as well as intellectually. (The
best and most comprehensive is Eliška Fučíková's, a vast book that
weighs about the same as a medium-sized child.) Demonstrating
the sheer scale of Rudolf's cultural project involves an endless
parade of artists, astronomers and adepts, each with their own

complex interconnections and smorgasbord of esoteric interests. It is no wonder Rudolf II was such a disastrous political leader – all his energies were channelled into commissioning artists, adding to his collections and managing the kaleidoscopic world he was creating inside the precincts of Prague Castle. There was no time left for dealing with foreign, domestic or religious policy, and little appetite for it, as his mental state became increasingly fragile. During the last years of his life, he was attended by just a handful of servants and his doctor, who were not to speak to him unless it was absolutely necessary. He no longer left the safety of the castle and died there in 1612, surrounded by the precious objects he had amassed.

Anyone imaging a turreted Disney affair or a bleak stone fortress might be disappointed. Prague Castle is, in fact, a series of monumental palaces with a majestic Gothic cathedral in the centre, on a hill overlooking the city. It would take several visits to take in all the sights and exhibits on show today. In Rudolf's day, it was a town within a town, with 'countless shops' and 'a terrace two hundred paces in length' specially reserved for fighting duels.[1] There were workshops, stables, gardens, palazzos belonging to Bohemian nobles and huge galleries housing the ever-increasing imperial collections. The view over the city is breathtaking; the red-tiled roofs tumble into the distance fringed by clouds of verdant green trees and parks while the river curves gently by with its succession of bridges and shallow weir.

Today, the city's architecture is a joyful riot of Gothic, Baroque, Neo Classical and Art Deco painted every colour under the sun and decorated with sculptures, flourishes, carvings and gilding; 'more is more' could well be its motto. Charming and garish, ancient, austere, ornate – it's all here in a glorious, heady melting pot. It was a very different place in 1591, when the English traveller Fynes Moryson visited on one of his long itineraries around Europe. 'The streets are filthy, there be diuers large marketplaces, the building of some houses is of free stone, but the most part are of timber and clay, and are built with little beauty or Art,' was his brusque verdict.[2] The classic Englishman abroad, he goes on to note approvingly that he managed to find some 'English Oysters pickeld' – expensive, but worth it for a taste of home.

The small towns that collectively made up sixteenth-century Prague are still discernible today: the castle on the crown of the hill with Lesser Town (Mala Strana) sprawling below it, and Old Town and New Town on the far bank of the Vltava river (known as the Moldau in German), reached via the iconic Charles Bridge, whose foundation stone was laid by the Holy Roman Emperor Charles IV at 5.31 on 9 July 1357, chosen for the palindrome 1357 9/7 5.31 which would, God willing, ensure the bridge's longevity. In this instance the invocation worked – the bridge still stands and for centuries was the only crossing point.

The Jewish ghetto is also on this side of the river. First settled in the twelfth century, only a few synagogues survive today, along with the cemetery and town hall. The Hebrew scholars who lived there during Rudolf II's reign brought the study of the Kabbala to his court through their translations and manuscripts. Now elevated to the official capital, the city was booming as thousands flocked to seek their destiny on the periphery of the Habsburg court and the streets, choked with mud and manure, thronged with quacks, prostitutes, beggars, preachers, thieves and charlatans – all on the lookout for opportunity, for scraps of fortune falling from the imperial household, high above them on the castle mount.

As the eldest son of Maximilian, Rudolf was brought up to rule, although primogeniture was not a given. He was educated from an early age in languages, mathematics and history by a selection of tutors chosen by his cultured father, whose favourite among them was the gloriously named Ogier Ghislain de Busbecq. Busbecq was the imperial envoy to the court of Suleiman the Magnificent in Constantinople, a delicate role that required sophisticated diplomatic skills. He is celebrated (perhaps wrongly) for bringing the very first tulip to Europe, along with lilac trees and the snowflake amaryllis, to the imperial gardens in Vienna from Turkey. From here Carolus Clusius took some tulips to Leiden to the botanical garden where they grew well, and within decades Holland was in the grip of a mania the Dutch have never recovered from.

When Rudolf was eleven years old, and his brother Ernst was ten, their mother insisted they were sent to be educated by her brother Philip II of Spain and imbued with the fanatical Catholic spirit of his court. She was also keen to get them away from the (in her view) diabolical Protestant influences that had corrupted their father and so many of his subjects in central Europe. So in 1563 the two little boys set off with their tutors on the long journey over the Alps. When they returned eight years later, they were young men, changed forever by their experiences in Spain. Their father, Maximilian II, was alarmed by their stiff manners and Philip Sidney, visiting Prague in the mid-1570s, noted that Rudolf was 'sullein in disposition, very secrete and resolute...extreemley Spaniolated'.[3]

Rudolf never shook off this formality; combined with his natural shyness which developed into serious introversion as he aged, it made him appear haughty and rude. He was obsessed with his own status but could be curiously informal with scholars. Both

Dee and Brahe were welcomed with private audiences alone with the emperor, while ambassadors were often kept waiting for months to be admitted into the imperial presence. This enhanced Rudolf's reputation for eccentricity, adding to the store of ammunition his brother Matthias and his other enemies were able to gather against him. One of their greatest complaints concerned his religious policy. Maximilian II had adopted a relaxed, tolerant, yet engaged, approach to the religious groups in his lands. A liberal, intelligent man, he was deeply influenced by the Lutheranism professed by so many of his subjects. Rudolf was similar – open-minded and happy for Protestants, Catholics and Jews to mingle at his court and in his cities – but he was less interested than his father. For him, religion was extremely troubling and he did his best to avoid dealing with it as much as he could, a policy that both presaged and contributed to the horrifying conflicts that were to come. He was blamed by some for the brutal Thirty Years' War into which the empire dissolved after his death.*

In spite of speaking several languages, Rudolf was a man of few words; hardly anything he wrote himself survives and the letters that do are official and offer no insights into his personality. We can only glimpse him in the reports of others, the diplomats and visitors who attended him, the artists who painted him and the scholars who sought his patronage. He is a paradoxical figure: the most powerful man in Europe at the centre of the largest and most prolific court, but silent and elusive, concealed among his artworks, hidden in his laboratories. Many descriptions of him are coloured by the writer's bias or frustration with his political apathy and religious ambivalence. The people who knew him best, the scholars

---

*   Fought between 1618 and 1648, this was one of the worst conflicts Europe has ever witnessed, caused by the struggle for religious and political dominance of the continent.

he employed and worked alongside, did not tend to write about their experiences because so much of the work they were engaged in was, by its very nature, secret. Most sources agree on one thing, however: 'Rudolf was, in the strictest sense, extraordinary.'[4] Even in his last years, when his mental health had deteriorated significantly, a Tuscan diplomat recognised, 'The Emperor's amazing knowledge of all things, ripe judgement, and skill have made him famous, while his friendliness, steadfastness in religion, and moral integrity have won him popularity; these were the principles of his outstanding and remarkable reign which gained the plaudits of the whole world.'[5]

Determined to preserve the purity of their blood, the Habsburgs pursued a vigorous programme of intermarriage. Incestuous unions between first cousins (like Rudolf's parents) or uncles and nieces were commonplace, shrinking the Habsburg gene pool to a murky puddle. Succeeding generations were beset by deformity, insanity and an alarmingly large lower jaw – the tragic consequences of the inbreeding policy also evident in Rudolf's own fragile mental health and peculiar facial features. Despite this, Rudolf inherited many positive attributes from his Habsburg forebears, especially his father Maximilian II, who was both intelligent and intellectually curious, a true 'son of the Renaissance'. Maximilian spent time as a young man at various German courts; like so many of the other figures we have met, he was profoundly influenced by Philip Melanchthon's educational philosophy. More controversially as the Holy Roman emperor, he was interested in Lutheranism, which caused major issues with his Spanish cousins, and in particular his wife, Maria, the daughter of Charles V. Her fervent Catholic faith and passionate adherence to her Spanish upbringing caused endless conflict with Maximilian. The union was not a happy one.

Rudolf himself never married, and so failed in his principal duty of producing an heir (astonishingly, none of his five brothers had any sons either) and a new generation of Habsburgs to follow him. Negotiations for a union with Philip II's daughter, Isabella, began in 1568 and continued sporadically for over two decades as Rudolf weighed up the merits of the dowry with the prospect of a wife. Which territories Spain would hand over as part of the marriage settlement was another sticking point; Rudolf wanted Milan, among other things, and contemporary reports are full of the wrangling that went on behind the scenes. To Rudolf's fury, Isabella was finally wed in 1599, at the age of thirty-three, to his younger brother Albrecht. Perhaps his parents' miserable marriage, which nevertheless produced sixteen children, made it hard for Rudolf to commit. The parallels cannot have escaped him: a liberal Austrian Habsburg allying with a devout daughter of Catholic Spain.

Rudolf was crowned King of Bohemia in 1575 and the following year his father died unexpectedly. Rudolf was his sole heir and not long afterwards, the rulers of the numerous imperial states elected him emperor. His brothers were aggrieved – the generous allowances paid to them were no substitute for political power – setting in train the process that would eventually lead to his overthrow. Rudolf became increasingly keen to put some physical distance between himself and his family, who were often in Vienna. He began spending most of his time in Prague, and in 1583, officially made it his imperial capital. He had always loved the city, which was bigger than Vienna and had been the capital of the empire earlier in the century. Another attraction was that it was three hundred kilometres further away from the Ottoman forces, who were now in possession of much of the kingdom of Hungary.

Prague had been a centre of religious plurality long before Rudolf was born. Since the early fifteenth century it had been home to the proto-Protestant Hussite movement, for many years effectively cut off from its Catholic neighbours. As a result, the city and its institutions were isolated and depleted, ripe for the renewal Rudolf II brought with him in 1583.

One of the first things he did was renovate and improve the large castle complex, continuing his father's projects and adding new ones of his own like installing laboratories and workshops, planning herb gardens and building a new wing to house the growing collections. Maximilian II had built a ball-games hall and a real tennis court, a shooting range, aviaries, a menagerie, a summer house and a fantastic bronze fountain. To these, Rudolf added vast stables, a new palace, covered walkways connecting the main buildings and a chapel on the western wall of St Vitus cathedral. According to one visitor, 'Inside the castle is a stable, which must be one of the best equipped anywhere, because there are always about three hundred horses from all possible countries and they are among the finest in the world.'[6] Rudolf adored horses, lavishing them with affection in a way he found impossible with human beings. Around the edge of the castle grounds was the Stag Moat, a huge park established earlier in the century that was home to deer and other animals for hunting.

Collecting was in Rudolf's blood. No family matched the Habsburgs as patrons in the sixteenth century, even in the Italian city states. This inheritance took staggering material form. On his father's death, Rudolf became custodian of the imperial library, thousands of works of art, precious stones and natural curiosities accumulated by his forebears. In the Middle Ages, royal collections were essentially treasuries of precious metals and jewels. During the Renaissance period, they expanded beyond all recognition to contain anything the ruler desired, becoming known as *Kunstkammer* or cabinets of curiosity. Rudolf's included a living dodo, a unicorn horn, thirty-seven cabinets of minerals, and a 'silver cylinder covered with crystalline glass' which may well have been one of the first telescopes in existence.[7]

As collecting became a fundamental attribute of kingship, so did displaying the treasures, demonstrating the ruler's taste and knowledge, and his ability to procure the rare and the priceless. Rudolf's collections were also an attempt to reassemble the fractured universe and discover its secrets, a microcosm of the world of knowledge he was so eager to explore, with multiple layers of value: financial, spiritual, prestige, practical, rarity, intellectual and talismanic. The boundaries between natural and artificial objects were blurred; there were art works made from natural materials like polished stones, exotic seeds made into drinking vessels and plants intricately recreated in silver and gold. Scholars were encouraged to study the precious objects, to use them to test and correct the fabulous theories inherited from medieval bestiaries. Someone had a close look at the unicorn horn and decided it was either from a rhinoceros or a narwhal, leading to the disappointing conclusion that unicorns did not, in fact, exist. This proto-scientific approach was new, and transformative.

Alongside the fantastic collections, Maximilian bequeathed his son a vibrant community of people. He employed some of the most inventive and interesting minds in Europe and when he died, many of them remained at the imperial court. Carolus Clusius, the botanist who worked closely with Wilhelm IV, was called to Vienna in 1573 to be Maximilian's herbalist and plant-collector, a post he held for fifteen years. He had studied with Melanchthon in Wittenberg so knew Peucer and published books with Plantin in Antwerp, and he was close to the German physician Johannes Crato, another friend of Melanchthon who had lived with Martin Luther while studying at the University of Wittenberg.

Crato was born in Wrocław in south-western Poland, a major hub on the sixteenth-century map of knowledge, with an excellent school and, as we have already seen, sometime home of Andreas Dudith, Paul Wittich and John Craig. After studying in Padua, thanks to the support of Wrocław town council, Philip Melanchthon and Joachim Camerarius, Crato was called to the imperial court by Ferdinand I in 1560 and appointed as his personal physician. He also went on to serve Maximilian, who showered him with rewards and made him a count. He left court after Maximilian died, having performed the first recorded autopsy on his body, but was soon recalled to tend to Rudolf, who he served until 1581. Crato was an influential figure at court and in the Republic of Letters, respected and valued for his medical ability, his towering intellect and many friendships.

The most influential of Rudolf's doctors was Tadeas Hajek, who had served both Maximilian II and Ferdinand I, born in Prague into a prosperous local family. Like so many of the figures we have encountered, he worked as a doctor, but was also a talented astronomer and astrologer, with a passion for alchemy. He was Rudolf's top intellectual recruiter, a prolific letter writer responsible for attracting many other scholars to the city. He was close to Sir Philip Sidney, under whose aegis he sent his three sons to England to be educated, and wrote on several subjects: the new star of 1572, various aspects of medicine, astrology (which he believed in passionately) and brewing techniques – a peculiarly Czech interest.

Sometimes it is difficult to take in the myriad connections of the sixteenth-century world of learning. Everyone, it seems, knew everyone, and almost all of them visited Prague at some point. Despite spanning an entire continent, this community was close-knit and vigorous, its members in constant contact with one another, arguing, praising, promoting. Of course, there were many examples of professional envy and malign behaviour (Brahe and Ursus spring to mind) but in general there seems to have been an atmosphere of positivity and support, especially when it came to encouraging the younger generation. Rudolf was at the very apex of this community in more ways than one: as the leader of the intellectual world in this period, and also as the most generous, supportive patron of the age, possibly any age.

Paintings made up a huge part of the Prague *Kunstkammer*. By the time of his death in 1612, Rudolf owned over 3,000. He commissioned hundreds, and many were made in the castle studios by the artists who lived there at his command. Together, this group of painters and sculptors, who originally came from Italy, the Netherlands and Germany, took the Mannerist style to its peak, producing masterpieces that were often shockingly erotic, imbued with a louche sensuality that added to Rudolf's dissolute reputation. They took the realism of Renaissance art and combined it with Neoplatonist ideals and the occult to create new fantastical, uniquely Bohemian, visions of the world. Symbolism and mystery took centre stage: nothing was quite what it seemed.

Philip II was a major inspiration for Rudolf's artistic patronage. As a young man at the Spanish court, he had fallen in love with the works of Albrecht Dürer and Pieter Breugel the Elder in the majestic royal palaces of Madrid. Seeking out their work, and especially items that had been commissioned by his ancestors, was an important strand in his collecting. In 1606, after lengthy negotiations and huge sums of money, Dürer's painting *The Feast of the Rosary* arrived at the *kunstkammer* in Prague. It had been carried over the

Alps from Venice, eased out of the altarpiece of San Bartolomeo, the church near the Fondaco dei Tedeschi, headquarters of the German Merchants Fraternity who, led by none other than Jakob Fugger, had commissioned the painting.

Apart from the glowing figure of the Virgin Mary in the centre, two figures would have stood out to Rudolf as he stood admiring his latest purchase. On the right hand of the Virgin, wearing a voluminous red robe, is his great-great-great-grandfather, Frederick III, painted with the likeness of his son, Maximilian I. Behind him, leaning against a tree and staring boldly out at us, is Dürer himself, his luxuriant red-gold curls blending into his apricot-hued, fox-trimmed gown. In his hands is a parchment, inscribed 'Albertus Dürer Germanus' with his AD monogram, the master of stylish self-promotion looking like a prince and reminding everyone he is not just any old Venetian painter, but something far more precious and rare. Would gazing at this painting have given Rudolf a feeling of connection with his Habsburg heritage? A feeling that he was carrying on an important tradition, that, in this regard at least, he was fulfilling his destiny as Holy Roman emperor, no matter what his brothers and other critics said? Or did it just make him wish he had the talents of Albrecht Dürer at his disposal?

In their multifaceted roles as *kunstkammer* curators, Rudolf's artists had to be flexible, imaginative and willing to learn new skills. Many of them lived in the castle complex and worked in the studios provided, collaborating and inspiring each other in an atmosphere of creativity and openness. This was a unique environment in terms of scale; there were of course many other courts where artists were employed in numbers and had the chance to work together, but none were near the size of Prague, nor did they have the same collections to use for stimulation and study. Dürer's detailed studies of animals and plants revealed the aesthetic and intellectual value in depicting nature with the same level of dedication as religious images and would later develop into the genre of scientific illustration. We have already seen it used by Wilhelm IV when identifying plant specimens, encouraged no doubt by Carolus Clusius and Joachim Camerarius the Younger who both worked at the imperial court.

In art, the genre of still life was in its early stages in the Netherlands. Fruit and flowering plants were the most popular subject matter, but dead animals, shells and other natural items also featured – many of which would have been in *kunstkammer* collections. In 1590, Rudolf appointed Joris Hoefnagel as a court painter. Originally from Antwerp, he was not a professional artist but a typical man of the Renaissance: a poet, a traveller, educated, intelligent, interested in everything. He is most famous for his illustrations in the bestseller, *Civitates orbis terrarum*, a sort of early modern *Lonely Planet Guide* which described European cities. Hoefnagel was a miniaturist and produced tiny studies of shells, flowers and butterflies for Rudolf using the *kunstkammer* as his prop box.

Rudolf also loved larger studies of nature, like those he saw in the Pieter Bruegel the Elder's paintings in Spain. He recognised Bruegel's desire to depict and understand the natural world in his own quest for knowledge and felt a deep connection with him. He employed various artists to continue this tradition; they created faithful images of the surrounding Bohemian countryside and more imaginative scenes inspired by their travels in Italy. In 1604 Bruegel's son Jan (also the Elder) visited Prague. He never knew his own father, who died when he was just a year old, but we can safely assume Rudolf questioned him about Pieter's work and worldview. He did not stay long in Bohemia, and returned to Antwerp, where he bought a house called The Mermaid and began working for Rudolf's brother Albrecht, now governor general of the Netherlands.

Alongside the artists' studios in the castle, Rudolf built workshops for the many other craftsmen he employed – the Miseroni family, gem cutters from Milan, set one up that outlived their patron and continued to flourish into the 1620s. Rudolf invested heavily in the search for precious metals and stones, an intrinsic element of his passion for alchemy and a huge source of income for the imperial coffers. He often visited the workshops and loved sitting on the workbench beside his artisans, experiencing the magic of carving metal, heating minerals or planning the composition of a painting. The Tuscan ambassador was not impressed by this: 'For he himself tries alchemical experiments, and he himself is busily engaged in making clocks, which is against the decorum of a prince. He has transferred his seat from the imperial throne to the workshop stool.'[8] Most of the work went on within the castle, but Rudolf also purchased a property on the outskirts of the city, where he built a sawmill and developed it into a sort of craft centre-cum-refuge for himself with a grotto, gallery and fishponds alongside a smithy, glassworks and a workshop.

Alchemy was one of the biggest enterprises at the Prague court. Rudolf is said to have employed around 200 alchemists, some based at the castle, others elsewhere in the city. There is no trace of their furnaces or stills in the castle today, but there is an intriguing place on the other side of the river, called the Museum of Alchemy.[9] In the vaults below are the reconstructed remains of alchemical laboratories. With ovens built into the walls and a glass furnace, they have been dressed with distillation vessels and other paraphernalia so they look as if the alchemists have just stepped out for their dinner and will be back to relight the fires soon. Upstairs in the old-fashioned apothecary-style shop, the shelves are decorated with bunches of dried herbs and scrolls, and ornate painted bottles of the elixir of eternal youth are for sale, promising to prolong life and bring good health. The elixir has been concocted using seventy-seven different herbs, based on the same recipe Tadeas Hajek used to make for Rudolf II, although the modern version does not contain opium. It feels like an outpost of the Harry Potter universe, an atmospheric glimpse into the lost realm of early modern alchemy.

Prague is a spooky place, full of ghosts and magic, something the tourist board capitalises on today. It's not always clear where reality ends and fantasy begins – the boundaries are indistinct, just as they were in the sixteenth century. Many of the people who claimed to possess magical or astrological powers were 'puffers' – con artists doing a roaring trade duping the masses. They travelled from town to town, moving on to the next place before they were found out. There's no doubt a few of the people Rudolf employed fell into this category; alchemists, even the well-educated ones, generally didn't tend to stay long in their posts.

That shouldn't overshadow the important, proto-scientific work that also went on in early modern laboratories, however. Alchemy in this period was a wide-ranging activity that resulted in many important discoveries. Rudolf's alchemists studied and used the minerals in his collections for experiments; this was the great age of mining exploration and pushing the boundaries of knowledge in metalworking and metallurgy. Like Wilhelm IV, Rudolf's interest in alchemy did not stem from the obsession with turning base metals into gold. His aims were altogether loftier and had more to do with learning about the properties of metals and minerals, working towards discovering fundamental secrets about how the world worked.

*
* * *
*

Soon after Rudolf established Prague as capital, two familiar would-be alchemists hove into view. After a disappointing stay in Poland, John Dee and Edward Kelly, urged on by the angels, decided to try their luck at the imperial capital, leaving their wives and children back in Kraków for the time being. They arrived in August and stayed at a house 'by Bethlehem in Old Prague' (a chapel where Jan Hus had preached) loaned to them by Tadeas Hajek, friendly with Dee since they had met at Maximilian II's coronation in 1563. Dee was pleased to find 'very many hiero-glyphical notes philosophical, in birds, fishes, flowers, fruits, leaves and six vessels' in the study, and an alchemical verse written on the wall.[10] It was the perfect setting for their own intellectual exercises, which by this point predominantly involved Dee witnessing and recording Kelly's conversations with angels and other spiritual creatures. The priority was to get an audience with Rudolf, so Dee wrote a letter which he gave to the Spanish ambassador to pass on. A few days passed; Kelly's behaviour was as erratic as ever. One evening at Hajek's own house, he got 'marvellously drunk' and, 'with wine overcome', threatened to cut off another fellow's head.[11] Dee spent much of the next day calming the situation down.

On 3 September a message arrived summoning Dee to the castle. He hurried over Charles Bridge and up the steep streets of Mala Strana. Presenting himself at the castle, he was led through several anterooms before finding himself alone in a privy chamber with Rudolf, who was sitting at a table with his letters and a copy of the *Monas Hieroglyphica*, dedicated to his father, Maximilian II. The emperor admitted that this book had been beyond him (just as it had been for Elizabeth I and doubtless many others) and said he had heard that Dee had something important to tell him. Of that, there was no doubt. Dee launched into a lengthy explanation that, after forty years spent in study, he had concluded that the only route to higher wisdom was through communing with the angels. Moreover, he continued, without a trace of hesitation, those very angels had told him that they were displeased with Rudolf, but 'If you will hear me, and believe me, you shall triumph. If you will not hear me, the Lord…will throw you headlong down from your seat.'[12]

This was an unbelievably risky thing to say to anyone, still less one of the most powerful men in the world who was well known for his aversion to religion. Sounding more like a deranged Jehovah's Witness than an erudite alchemist, and certainly nothing like an imperial philosopher and mathematician (the position he later requested), Dee was so convinced he had a direct line to God via Kelly, he gave no thought to how his message might sound. As soon as he could, Rudolf ended the meeting, saying, in his almost inaudible voice, that he would hear more another time. Dee never saw him again.

After this incident, all communication was carried out through an intermediary called Dr Curtz, who Rudolf charged to listen to Dee and then write brief reports for him. Rudolf had clearly seen through Dee's claims to be able to make the philosopher's stone and his melodramatic predictions; it was hard to move in central Europe in this period without bumping into someone proclaiming the imminence of Judgement Day, heralded by the great conjunction of Saturn and Jupiter predicted for that very year and culminating in 1588.

At some point that autumn, Jane Dee and the rest of the household arrived from Kraków. We can only imagine how she felt on reaching Prague. She had now been travelling almost constantly for well over a year, she had been seriously ill and she was pregnant again – Michael Dee was baptised in St Vitus cathedral in early spring 1585. How many times in the those years on the road must Jane have longed for her home and hearth on the banks of the Thames? The filthy weather, foreign food, bumpy roads, flea-ridden beds and the myriad other discomforts she must have had to endure, while worrying about her children's health, her husband's prospects and Kelly's machinations, don't bear thinking about. She sent Dee letters when they were apart, writing from Prague in February 1587, 'Sweetheart, I commend me unto you, hoping in God that you are in good health, as I and my children with all my household am here, I praise God for it.'[13] Money was a constant worry and Kelly's behaviour and 'visions' were increasingly hazardous. On 21 March, Jane wrote a petition for Kelly to pass on to the angels. A rare instance where we hear her voice: 'We desire God of his great and infinite mercies to grant us the help of his heavenly ministers…of sufficient and needful provision for meat and drink for us and our family, wherewith we stand at this instant much oppressed… I Jane Dee humbly request this thing of God, acknowledging myself his servant and handmaiden; to whom I commit my body and soul.'

The spirit responded, 'I cannot give thee that thou desirest…so I counsel thee: let thy husband arise and gird himself together, and…hasten out of this place: for my thinks they dissemble.'[14] They were quite right. Dee's insensitivity, his inability to judge character and read situations, and his indiscretion about the angel conferences were causing dangerous rumours that he was 'A bankrupt alchemist, a Conjuror, and Necromantist'. Things went from bad to worse, but Dee neither saw through Kelly nor recognised his own shortcomings. 'I became hindered and crossed to perform my dutiful and chief desire; and that, by the fine and most subtle plots laid,' he claimed – his failures were always the fault of others.[15]

Events came to a head at the end of May 1586, when, under pressure from the papal nuncio who suspected heresy, Rudolf banished them. Dee went to Kassel, but was disappointed by his visit, although Wilhelm was supportive, and they corresponded occasionally afterwards. By this point their interests had diverged. Dee's obsession with divine revelation and his belief in the philosopher's stone (which he had in the form of a red powder), were not popular at Wilhelm IV's court, where they were busy making observations for the new star catalogue.

Dee's Bohemian adventure did not end there. Having spent the summer trailing around southern Germany, he and Kelly were taken into the protection of the Bohemian noble Vilém Rožmberk (William of Rosenberg), a keen alchemist and enthusiastic patron of learning. At Rožmberk's Třeboň estate, the angelic conferences continued intermittently. Dee records in his diary that Kelly 'sent for me to his laboratory over the gate: to see how he distilled sericon [antimony]' – the mercurial scryer was reinventing himself as an alchemist.[16] As Kelly's reputation grew, Dee's diminished. Realising this, Kelly began excluding his former master from the alchemical processes, making it clear to Rožmberk that he was surplus to requirements.

Whatever Kelly was doing in the laboratory, Rožmberk was impressed. He sent him to Prague, where he made 'a public demonstration of the philosophers' stone', apparently transmuting a small amount of mercury into gold.[17] When news reached England, Cecil and Elizabeth I began trying to persuade him to return; the idea that an English alchemist was in Bohemia gaily transmuting piles of gold for another monarch was not acceptable. Dee on the other hand received no such requests, but having run out of options on the Continent, he had no choice but to plan his journey back to England. In February 1589, just before his departure, he 'delivered to Mr Kelly the powder, the books, the glass and the bone for the L. Rosenberg', effectively handing over what remained of his alchemical power.[18] Kelly was borne off to Prague by his patron and made a baron, while Dee and his family began their miserable, and ruinously expensive, journey home. They sailed into Gravesend on 22 November 1589, six years and one month almost to the day after their departure. Dee returned home to find that his library had been ransacked and many of his instruments taken or damaged. The house at Mortlake never regained its former position, and Dee's fortunes continued their downward trajectory. He died there in penury in 1608 or 1609.

*   *
  *
   *

Dee's impact on the intellectual scene in Prague was negligible, but a postscript remains in the form of the enigmatic Voynich manuscript. Written in a cypher that has never been decoded, a seventeenth-century letter inside states that Rudolf II purchased it from John Dee for the enormous sum of 600 ducats in October 1586, believing it to contain the work of Roger Bacon. 'The Most Mysterious Manuscript in the World' is now housed in Yale University Library but its drawings of astrological symbols, plants and naked female figures continue to elude our understanding. As such it serves as an apt allegory for the recondite world of early modern alchemy.

Rudolf owned a great many occult texts and doubtless kept them close to hand in Prague Castle where they could be consulted with ease. However, most of the imperial library was still in Vienna under the careful custodianship of Hugo Blotius, the Dutch scholar who had been appointed to the post by Maximilian II in 1575. He and Rudolf worked together to increase the collections, keeping a lookout for other libraries to purchase, like that of the Hungarian Joannes Sambucus, yet another of Melanchthon's pupils, who worked at the imperial court as historiographer. On his death in 1584, Rudolf and Blotius made sure this collection, one of the greatest in Europe, was absorbed into the imperial library. The scientific holdings were second to none and read like a who's who of astronomy, astrology and alchemy: Copernicus, Cardano, Peuerbach, Frisius, Sacrobosco, Postel and della Porta, alongside all the great classical names, and an impressive number of works by the giants of Arabic scholarship. In amongst the books, mentioned specifically by Sambucus in his will, was the treasured but later discredited unicorn horn. Oswald Croll, one of Rudolf's physicians from 1602, used the collections when researching his *Basilica Chymica* of 1608, which explores the relationship between the microcosm and the macrocosm, a popular idea at the time. The second half is about pharmacology – it became the most popular text on the subject and a classic early modern combination of the 'scientific' with the mystical that features again and again in the period.

It is hard to ascertain exactly where Croll and his colleagues would have worked in the castle, and how they used the collections. Was there a system for borrowing items and taking them down to the workshops? Or was there no separation between where the treasures were kept and studied? Pierre Bergeron, who visited in the early seventeenth century, described the astronomical instruments thus:

> On the ground floor arcade, in the galleries of the summer palace, one can see countless spheres, globes, astrolabes, quadrants, and thousands of other mathematical instruments, all made from bronze and tin and fantastic in size. There are analemas, quadrants, spheres, dioptries, and Ptolemaic scales for the exact determination of height, distance, and constellations of the sun and

stars. They are divided into many smaller parts and are on a scale of sixty. There are also many instruments for the measurement of weight.[19]

There were many instruments in Rudolf's collections, and his astronomers presumably took them outside and used them to make observations. There were also the castle workshops, where anything you desired could be made. Among the craftsmen on hand were Jost Bürgi, still nominally living in Kassel, but coming to Prague regularly and making commissions for the emperor. Their relationship began in 1592 when Bürgi, in the city to deliver one of his sought-after mechanical globes, presented Rudolf a manuscript showing his new rapid method of computing sine functions, an accurate sine table for every minute from 0° to 90°. Impressed, Rudolf had been trying to tempt him to settle in Prague ever since and was very pleased when he was finally prised away from Kassel in 1604. The star-crossed alliance of Bürgi's technical genius with Kepler's intellectual brilliance would light up the Prague firmament in the following decade. Bürgi's workshop was in Vikářská ulička until the 1620s when he left Prague. Johannes Kepler used a sextant he made, and it was still in the *kunstkammer* in 1782 when it was sold as part of the Josephine auction.

Bürgi had stepped into the shoes of another technical wizard – Erasmus Habermel. Habermel had lived and worked in the castle from around 1576 until he died in 1606, and been married to Susanna Solis, sister-in-law of the painter Virgil, whom he probably collaborated with. They lived in the castle precinct, and she had a shop selling clocks, part of the thriving creative community Rudolf had created. One of the sextants Habermel made for Rudolf is on show in the Prague Technical Museum – it is an object of quite breathtaking beauty and elegance, all the more miraculous because it survived the chaos of the years after Rudolf's death and has been preserved for posterity.

Rudolf's instruments were about to be joined by the most impressive collection of the time, those belonging to Tycho Brahe. When we left him, Tycho was still on Hven, surrounded by his instruments and assistants. In the years that followed things went spectacularly wrong, starting in 1588 when Frederick II died. The new king, Christian IV, did not appear either very interested in astronomy or impressed with the court astronomer. Without his royal sponsor, Tycho was exposed to resentment that had been building up over the years, anger over the amount of money the state was pouring into his pockets, coupled with his arrogant neglect of his responsibilities. As he gained his majority, Christian IV was keen to impose his own stamp on the country – Tycho would be the lightning rod. It became clear he could not stay in Denmark, so in 1597 he boarded a vessel bound for Rostock, with as many of his books and instruments as he could carry. A friend, Count Rantzau, offered Tycho a safe haven at Wandsbeck, his estate in Holstein. Two years later, after much to-ing and fro-ing, Tycho set forth for Bohemia.

He entered the city in the midsummer of 1599 and was welcomed with warmth and enthusiasm. Before he had even met the emperor, the imperial private secretary Lord Johannes Barwitz took him to see a palace just to the west of the castle, 'built in the Italian style with beautiful grounds', which had belonged to the long-suffering and now deceased Dr Curtz. It was his, should he want it. Tycho admired the property, but the tower Curtz had built for making observations was demonstrably far too small, and 'would scarcely suffice for a single one of my instruments', so Barwitz offered him the choice between three estates outside the city instead. A few days later, Tycho was summoned to the castle. Like Dee, he was surprised to find himself in a chamber with the emperor, 'completely alone in the whole room without even an attending page'. Even more astonishing, Rudolf proceeded to call him forward and reached out his hand. Tycho presented some documents of recommendation, and the emperor welcomed him and promised to support him, 'all the while smiling in the most kindly way so that his whole face beamed with benevolence'.[20] Tycho struggled to hear everything because Rudolf was so softly spoken.

When the audience was over and Tycho was back in the ante-chamber, Barwitz, having spoken briefly with the emperor, asked if he could see the mechanical device on his carriage. Apparently Rudolf had spotted it while watching them arrive through the window. Tycho sent his son, Tyge, down to get it and explained to Barwitz how it worked so he could show his majesty. When the device was returned, Barwitz said that Rudolf would have a copy of it made by one of his astronomers. This rare first-hand report takes us right into the room with Rudolf, and provides the touching image of him excitedly looking out of the window to catch a glimpse of the famous astronomer he was going to host and support.

And support he did. Rudolf persuaded the parliament to grant Tycho a huge annual stipend of 3,000 guldens (more than many long-serving courtiers), back-dated to May 1598, when he had initially been invited to Prague, in addition to relocation costs. Moreover, Tycho moved into the Curtz house *and* took on the Benatky estate, six hours from Prague, which he immediately began rebuilding to suit his purposes. This was patronage on an industrial scale, and a potent recognition of Tycho's status as a nobleman, but even more so as the most successful astronomer in

Europe with his own system of the universe, his astrological expertise and stable of unrivalled instruments. The Brahe brand was never going to come cheap; it was just a shame it was so short-lived.

Much to his displeasure, Tycho had had to leave his four largest instruments behind on Hven; he sent his son Tyge back to fetch them in 1600. They were held up in Rostock and then in Hamburg and didn't reach Prague until October 1600, shortly after the ones he had left behind in Magdeburg finally turned up. Rudolf intervened but the local authorities seemed unwilling to part with them, until Tycho suggested they could earn some extra money by taking some barrels of Bohemian wine back as cargo.

Rudolf's support knew no bounds, and Tycho could not praise him enough. 'His Majesty not only shows me all goodness and does not allow me to lack anything but has also graciously shown a personal, fatherly concern for me…sends me game from time to time, harts, hinds, and wild boars, and also various living fresh fish to the kitchen,' he wrote in a letter to his sister Sophie, dated 21 March 1600. He went on to explain that, once his instruments arrived, 'His Majesty intends to come here to the castle [Benatky] to see them and my other work, both distillation and other things, for His Majesty has a strong pleasure in all such things.' He tells her about 'the buildings that His Majesty is having built for me here, the one a special observatory in which all of my instruments can be set up properly and in order, so that each will have its enclosed room to stand in, and the other a special laboratory'.[21]

If all this sounded too good to be true to Sophie, she was right. Rudolf, like monarchs throughout history, sometimes made promises he couldn't keep. A couple of months later, a visitor described thirteen rooms, each with a large instrument inside and no fewer than twelve assistants to operate them. They made rough observations of all five planets, but the focus of Tycho's work was preparing the observational data, collected over thirty years, for publication, and for that he needed skilled mathematicians. His first choice was his former assistant Longomontanus, who was reluctantly persuaded to accompany Tyge, the large instruments and several barrels of salted fish from Denmark to Prague. Other recipients of Tycho's entreaties included Johannes Müller and David Fabricius; he even toyed with the possibility of tempting Christoph Rothmann out of retirement.

The fates had other plans, however, and on 5 February 1600 Johannes Kepler rode through the gates of Benatky. This was the first time the two men had met in person, and Kepler must have been nervous. He had come to Tycho's attention in 1598 when he had unwittingly wandered into the middle of the ongoing (and vicious) controversy with Ursus.

When we last encountered him, Ursus was in Kassel showing off his universal system to Bürgi, Rothmann and Wilhelm. He published it as his own work in 1588, further enraging Tycho, who spent the rest of his life trying to prove it was plagiarism in a very public, very ugly war of words. His anger was only increased when Ursus moved to Prague and Rudolf appointed him imperial mathematician. Kepler, young, inexperienced and keen to make contact with his superiors, sent copies of his book *The Mystery of the Universe* to several well-known astronomers, including Tycho and Ursus. He also wrote an ill-judged letter to Ursus, praising his talents and his universal system. Glad of any support, even from a nonentity, Ursus had published it, without asking, in a book containing vile attacks on Tycho and his family. Kepler's eyes were opened by his tutor Michael Maestlin and when he realised the situation, he was both mortified and furious. Of Ursus he said, 'I cannot believe that I expended so much sweat in praising him that the jackass can justly strut around.'[22] Apologising to Tycho (who he had also been keen to impress and make contact with) was his foremost concern, so he wrote a letter which Tycho received in Wittenberg, just before he set off for Prague. Tycho replied in a letter he sent to Gratz (now Graz), the town in Austria where Kepler had been living and working as a maths teacher. By the time it arrived, Kepler had already left Gratz and was heading for Prague. As a committed Lutheran, he knew he would not be able to keep his job as pro-Catholic forces in Austria moved to stamp out Protestantism as part of the Counter-Reformation, whereas in Prague under the protection of Rudolf II, he would be safe. Yet again, the vagaries of religious policy rippled through the astronomical community, uniting people who otherwise might have never met.

Kepler came from a noble family now somewhat on their uppers. His father was a mercenary so was largely absent during his childhood, and money was scarce. Fortunately, Johannes won a scholarship and was sent to a series of schools that educated young Lutherans for a career in the Church – another beneficiary of Melanchthon's reforms. From here he went on to the University of Tubingen. His grades are still in the archives: straight As across the board. One teacher commented that he had 'such a superior and magnificent mind that something special may be expected of him'.[23]

As he was nearing the end of his studies, he was put forward as a candidate for the maths teaching position in Gratz and, suddenly, an alternative path appeared, leading in a different direction to the pastorship he had been preparing for. Having carefully weighed his options, Johannes decided he would rather spend his time investigating nature than arguing about the relative merits of Calvinism and Lutheranism. He was also worried that his skinny frame and fragile health made him unsuitable for the Church – he had been born two months premature and survived various illnesses during childhood (including a bout of smallpox aged four that left him weakened and vulnerable to fevers for the rest of his life) – as in his view, a good vicar should be burly and have a good beard.

He set off in March 1594, arriving in Gratz on 11 April. Not many of the young noblemen were interested in maths, leaving Kepler with plenty of time to pursue his own ideas. He continued the research he had begun under the guidance of Michael Maestlin, his tutor in Tubingen, who had introduced him to Copernicus' *De Revolutionibus* and the debates surrounding it, and the work of Tycho Brahe. In 1597 he showed Maestlin the results – a book called *The Mystery of the Universe* which asked the fundamental question 'What is moving celestial bodies?' for the first time.

Kepler could ask this question because he was a convinced Copernican, the first to write a treatise grounded in the heliocentric system. He was also sure that the planets did not rotate on crystal spheres, which led him to wonder how they moved through space and what space was made of. He concluded that, whatever the force was that makes the planets move, it emanated from the sun, which made perfect sense if it was at the centre of the universe. He went on to explain that the spaces between the planets are defined by the five Platonic solids – as set out by Euclid: a pyramid, a cube, an eight-sided shape, one with twelve sides and one with twenty. He was thinking about the universe as a three-dimensional space where it had previously been expressed using two-dimensional models.

As far as he was concerned, Kepler had used geometry to see into the mind of God and reveal how he had created the universe, as he explained in a verse in the middle of the text:

But I look for the traces of Thy spirit out in the Universe,

I regard ecstatically the glory of the mighty celestial edifice,

This artful work, the mighty wonder of Thy almightiness.

I see how Thou hath determined the orbits according to fivefold
    norms,

With the Sun in the middle to donate life and light,

I can see that her laws regulate the course of the stars,

How the Moon achieves her transitions, suffers eclipses,

How Thou scatterest millions of stars across the realm of the
    skies.[24]

Maestlin was impressed. He organised publication in Tubingen, with the university's blessing. Kepler returned to Gratz, copies of his newly printed book in his luggage. After getting married a few months later, his salary was increased, his new wife got pregnant and everything seemed rosy.

Although he had generated some renown for the almanacs he had published over the previous six years, this new book was to be his springboard into the astronomical sphere, much as Tycho's work on the new star had been for him. If this was to happen, Kepler needed to make sure it got into the right hands, onto the right desks. He sent copies to several people and the list is instructive. Galileo Galilei, thirty-three years old and professor of maths at the University of Padua, received one. He hadn't published anything yet, but already had an international reputation as an inventor and engineer. In his response, he confided to Kepler that he was convinced by Copernicus but not willing to declare it publicly and asked for two more copies of the book. When they arrived, there was a letter in the parcel from Kepler urging him to be honest and open about his opinion and asking him to make some observations of the pole star and Ursa Minor with his superior instruments. As we already know, two other astronomers who received copies of the *Mystery of the Universe* were Tycho Brahe and Ursus.

When Kepler arrived at Benatky, any worries about how Tycho would receive him were soon dispelled. Far from holding him responsible for what had happened with Ursus, Tycho was thrilled to have Kepler on side, and lost no time in asking him to write a pamphlet condemning his hated rival. He did, however, treat his new assistant as just that, and this soon began to cause problems. Kepler protested that he hadn't been granted full access to Tycho's data and negotiations over the terms of his contract got very heated; at one point he even returned to Prague. Eventually everything was ironed out and Kepler set off to fetch his family from Gratz. By now it was summer, and things were finally taking shape at Benatky. Rudolf returned to Prague after several months in Pilsen, where he had gone to escape the plague, encouraged by Tycho in his role as astrological advisor. After such a long absence, there were many affairs of state to attend to and decisions to make, a task he could not possibly face without his imperial mathematician. Tycho was summoned; from now on, he would be living in Prague – Rudolf may have been a generous and attentive patron, but his needs trumped everything else, even the establishment of a new observatory.

When the instruments turned up a few months later, Rudolf told Tycho he could set them up on the second floor of the Belvedere summer house, built by his grandfather Ferdinand I near to the castle. He recorded observations on Christmas Night; however, most of his time and energy was spent drawing up forecasts for the emperor, consulting on military policy in the war against the Ottomans and the simmering religious tensions within the empire. Tycho and his household moved into the Curtz house he had rejected when he first arrived; it must have been a squash, especially once all of his instruments had arrived. Longomontanus returned to make his own life in Denmark, leaving the way clear for Kepler, who had spent that summer back in Gratz, working on optics. By October he was in Prague, where he discovered that Rudolf's legendary generosity had begun to dry up. Tycho was going to have to pay him with no extra help from the imperial coffers.

Kepler spent the next year, on Tycho's insistence, continuing to wage war on the now dead Ursus, and unsuccessfully trying to salvage some of his wife's property in Austria. When he got back to Prague in August 1601, Tycho took him to see the emperor. They made an agreement that they would publish a new set of astronomical tables named after Rudolf, with Johannes working as Tycho's assistant. The agreement happened just in time: a few weeks' later, Tycho was dead. His assistant inherited the project and the position of imperial mathematician (with a much-reduced salary). Most important of all, Kepler had access to 'the observations of thirty-eight years',[25] which Rudolf offered to purchase, along with the instruments, for the astonishing sum of 20,000 florins from the Brahe family. Needless to say, only a fraction of this amount was ever handed over.

Tycho's son-in-law and former assistant Frans Tengnagel had other ideas, however. He saw himself as Tycho's heir and persuaded Rudolf to appoint him as a second imperial mathematician, but his attempt to take over the data and the *Rudolfine Tables* project came to nothing in the end. For the time being, this forced Kepler to pursue other projects, with spectacular results. In 1604 he published *The Optical Part of Astronomy* which revolutionised and re-founded optical theory. Taking on the anatomy of the human eye, the nature of light and atmospheric refraction, it was on sale at the Frankfurt Fair in the autumn of that year.

Next, he turned his attention back to the project on the observations of Mars he had begun during his first month working for Tycho. As usual, he took a novel approach. He thought about how the earth would look when viewed from other planets, and how the constant motion of celestial bodies affected observations. Using his theory that the sun caused the physical motion of the planets, he realised that the traditional circular orbit needed to be squeezed to form an oval shape – an ellipse. This went on to become known as his first law of planetary motion (there were three in total), published in 1609, in one of the most important scientific books ever written: the *Astronomia Nova*, the New Astronomy. Rudolf had gathered some of the most brilliant people of the age in Prague, but Johannes Kepler was the brightest star.

Like Dee's, Kepler's family life was challenging. His wages as imperial mathematician were not overgenerous, and often they didn't materialise at all. He lived on the very edge of his finances, and struggled to support his children, especially after the death of his first wife Barbara in 1611. Taking a typically scientific approach to remarrying, he spent two years considering eleven different candidates before choosing Susanna Reuttinger. While this helped with life at home, it also meant the birth of several more children, although tragically, three of them died in childhood. His careful choice paid off; Susanna made him very happy and he wrote, 'she won me over with love, humble loyalty, economy of household, diligence, and the love she gave the step-children'.[26]

While things were difficult financially, Kepler flourished in Prague and made a great many friends among the scholars and artists employed at court. One of his closest companions was

Johannes Pistorius, the imperial physician who had taken over Hajek's post when the latter had died. Born into a Lutheran family in Hesse, he became a Calvinist before finally converting to Catholicism – experiencing the full spectrum of early modern Christianity. Professionally, he covered all the bases as a lawyer, a physician and a theologian, a man who shared his patron's love of the arts and sciences, and commitment to open-mindedness and toleration.

Despite a large age difference (Pistorius was twenty-five years older), he was very close to Kepler and the two loved nothing better than discussing the intellectual issues of the day. After a particularly passionate dispute about religion (Kepler was Lutheran), he concluded, 'I shall nevertheless remain your friend and servant because your mathematical talent and exceptional genius deserve as much.'[27] Thanks to Pistorius, we have a serious, personal record of Rudolf's mental state, an important counter-weight to the rumours his enemies put about. In his opinion, the emperor was melancholic but not obsessive, and at the mercy of various people around him who were playing on his fragility for their own ends. Among the nobility, another leading figure was Baron Ferdinand Hofmann von Grünpichl und Strechau (named for his godfather the emperor Ferdinand I). Educated in the Rožmberk household, he grew into a tolerant, learned man who amassed one of Europe's greatest libraries. His 4,000 volumes reflected the full gamut of the early modern intellectual world, a range similar to Dee's library in Mortlake, but Hofmann's huge wealth allowed him to have each one beautifully bound and orna-mented, something Dee could only dream of. Hofmann's collec-tion was a wonderful resource for his many scholarly friends who gathered on his country estates and in his house below Prague Castle to discuss and share ideas.

In 1611, Kepler presented another friend, Johann Matthaeus Wacker von Wackenfels, with a highly original New Year's gift. Walking home over Charles Bridge one freezing winter night, Kepler had noticed that the snowflakes falling around him were all hexagonal; he wrote down his ponderings on the phenomenon in a little book, 'On the Six-Cornered Snowflake', which he gave to Wacker in lieu of something precious and expensive, which he couldn't afford. Typically of Kepler, his musings turned out to be 'a perceptive, pioneering study of the regular arrangements and the close packing that are fundamental in crystallography'.[28]

Wacker played a significant role in the intellectual life of Prague in this period. He was another member of the Wrocław circle, close friends with Sidney, Dudith and Clusius; his job as a tutor had taken him across Europe when he was a young man, furnishing him with a wide circle of friends and contacts. Rudolf ennobled him, adding 'von Wackenfels' to his name, and he served as a councillor after settling in Prague in 1599, in a house just below the castle. He and Kepler shared many intellectual interests, not least astronomy itself, and apparently when Wacker heard about Galileo's discovery of the moons of Jupiter he raced over to Kepler's house in his coach, so overcome with excitement Kepler heard him shouting from the street. Kepler himself didn't use a telescope until 1610, when he borrowed one from the Elector of Bavaria, suggesting there was not one available in the Prague collections. However, we know that Dee gave one to Kelly before he left for England, and that it was then passed onto Rožmberk who presented it to Rudolf. Given the novelty of telescopes and corresponding variation in quality, perhaps this one was just not effective enough to be worth using.

Kepler took over imperial astrological duties after Tycho's death. Rudolf was especially concerned about the great conjunction of Jupiter and Saturn set for 1604, the last in an 800-year cycle which he believed could have cataclysmic consequences. The emperor had always been captivated by the power of astrology, avidly reading astrological books as a young man in the Escorial library and consulting the 238-page horoscope the legendary French astrologer Nostradamus had drawn up for him in 1564 when he was an impressionable, anxious twelve-year-old. A man with Rudolf's delicate mental state and limitless political power was bound to find solace in the idea that his destiny was written in the stars; this was buttressed with the appearance of the new star to coincide with his coronation in Hungary in 1572.

Kepler's position on astrology was more complicated. Reliant on the money he could make by writing calendars for a popular audience, he was sceptical about making predictions and the role of the houses of the zodiac, writing in 1601 that 'If astrologers do sometimes tell the truth, it ought to be attributed to luck.'[29] However, he did believe in the planets' influence on individuals and some 800 horoscopes he drew up survive in manuscript. He had to strike a balance between alerting Rudolf to the limitations of astrology while simultaneously being careful not to talk himself out of a job.

As the political and religious situation became increasingly fraught and complex, the emperor needed ever more reassurance. Matters came to a head in 1611, when Rudolf handed the crown of Bohemia to his brother Matthias, who had already taken the throne of Hungary several years before. Isolated and ill, Rudolf begged Kepler to remain with him in Prague. Nine months later, holed up in Prague Castle surrounded by his precious collections, Rudolf died, refusing the last rites on his deathbed. Kepler lost no time in packing his bags. He was embroiled in his own personal tragedy – a few months earlier, his wife had died of typhus and his son of smallpox, so he entrusted his two surviving children to a widow in Moravia and set off for Linz in Austria, where the position of provincial mathematician awaited him.

Matthias moved the court back to Vienna and most of the artists and scholars left to seek patronage elsewhere. A few stayed in imperial service and Bürgi remained in his Prague workshop making clocks. Without Rudolf, the mighty infrastructure of the world he created collapsed, like a circus tent when the central pole is removed. The long process of dispersing the vast collections began. The new emperor took the best items with him to Austria and his siblings helped themselves to whatever took their fancy.

In the chaotic years that followed, many things were sold to pay off debts and to finance the armies needed to fight the Thirty Years' War. In 1648, when the brutally effective Swedish army occupied the city, their queen, Christina, was disappointed not to find more treasures to take back to Scandinavia. Whatever was left in the abandoned state rooms was finally sold off in the Josephine Auction of 1792 – very few of the thousands of objects Rudolf amassed remain in the castle today. Despite this, his legacy is widely, if not commonly, recognised. In recent times historians have begun to look beyond his reputation for eccentricity and political failure. His achievements are now, rightly, celebrated. The transformative effect he had on the status of scholars and artists, many of whom he ennobled in recognition of their talents, helped to elevate painting and sculpture to the liberal arts and alchemy to a serious academic discipline. As a centre of learning, Prague was unprecedented in scale, breadth and ambition; it was an inspiration to succeeding generations, an example of what could be achieved with real investment in culture and science, the value of bringing diverse skills together and enabling collaboration.

The emperor was immortalised in 1627, when the *Rudolfine Tables* were finally published; the project that hung over Kepler for most of his career but which ultimately proved worth it for them both (and for Tycho, whose star catalogue was included) in terms of astronomical legacy. Recognising the 'immense superiority' of the *Tables*, astronomers long into the future used them as the foundation of their own work. They were the latest in a distinguished line stretching back through the Middle Ages to the ancient world, the stepping stones of astronomical progress.

When Rudolf died, the European world of learning lost its glittering epicentre. The artists and scholars dispersed, taking their brilliance with them, and his collections were scattered to the winds. In the following decades, political instability across the continent kept them moving, on the run from religious persecution and bloody conflict, with no safe haven of toleration and enlightenment to shelter in. The intellectual world fragmented, and it would be decades before a new centre rose to rival Prague.

7

ATLANTIS

A VISION OF THE FUTURE

Wonder is the seed of knowledge.

Francis Bacon

Our final destination cannot be found on any map. It can only be visited in the pages of a strange, slim novel written by Francis Bacon, First Viscount St Alban. *The New Atlantis* begins on a ship leaving Peru and sailing 'for China and Japan, by the South Sea'. Five months into this epic journey, 'there arose strong and great winds from the south', which blew the vessel into 'the midst of the greatest wilderness of waters in the world'. The crew, running low on supplies, had begun to give themselves up for dead when, suddenly, land appeared on the horizon.

This was the island of Bensalem, unknown, magical, a land populated by educated, civilised people who took the destitute mariners ashore and made them welcome. For Europeans this was literally the other side of the world, the furthest, most mysterious part of the planet, somewhere an island could easily be concealed amongst the unimaginable expanse of the Pacific Ocean. The location ensured Bensalem's separation from the rest of the world, and its people's isolation from the centuries of scholasticism Bacon was so opposed to.

Having ascertained they are also Christians, the locals welcome Bacon's mariners and look after them kindly. They explain the history of their strange island and its customs, and time passes in illuminating conversation. Then one day, an important visitor is announced, 'one of the Fathers of Salomon's House…clothed in a robe of fine black cloth, with wide sleeves and a cape…gloves that were curious, and set with stone; and shoes of peach-coloured velvet'. He arrives on a cedar chariot studded with crystal and gems, with an entourage of fifty young men in white satin coats and blue velvet shoes. Unlikely though it seems, this man is one of the directors of Bensalem's state-of-the-art research institute, which he goes on to describe to the stunned audience.[1]

In his vision of Salomon's House, Bacon addresses the problems that beset the centres of learning examined in this book. It is a call to arms for contemporary rulers to take scholarship seriously and invest in it accordingly, partly for its own sake but mainly to improve humanity's understanding of the natural world, something Bacon called 'dominion'. He describes in detail the complex hierarchy of Salomon's House, how the practitioners will work together and who will guide the hundreds of projects that are in progress. The island is a microcosm of the planet, with specialised areas for studying every feature: a cave system for experiments with refrigeration, special chambers and baths for medical study and therapy, gardens for the cultivation of food and medicines, but also for botanical study of every type, furnaces for alchemical investigation, perspective houses for looking at light and optics, sound-houses for music – an entire ecosystem in miniature.

In locating Bensalem in the New World, Bacon frees it from the baggage of old Europe. It represents a new start, pure and unsullied by the errors of the existing body of knowledge. This was a reaction to the discovery of the New World, encapsulated in the frontispiece image of Bacon's *Novum Organon*, which shows a galleon sailing off into the Atlantic and the unknown, while another ship passes it on its way home from the Americas, loaded with wonders. The motto underneath in Latin, 'many will travel and knowledge will be increased', says it all. This phrase was taken from the Book of Daniel and refers to the end of history – like Dee, Rudolf and others, Bacon was a firm believer that this momentous event was not far in the future, and that he was living through the final period of humanity, when lost knowledge would become accessible again. The ships are sailing through the Pillars of Hercules, gateway of the Mediterranean, the world of antiquity and ancient knowledge, which Bacon urges his readers to look beyond, outwards into the world, to discover new things and not be limited by old ideas.

It's not easy, beating a path through the vast thickets that have been written about Francis Bacon since his death, just as it is challenging to find a way through his own numerous titles. The iconic edition of his *Works*, exhaustively compiled by James Spedding in the nineteenth century, runs to a hefty fifteen volumes. *The Great Instauration*, his scheme for transforming scientific investigation, includes several different titles: *The Advancement of Learning, The New Atlantis, Sylva Sylvarum, Novum Organon* and the natural history of winds, life and death. Many more were planned but never written. Others, on lodestones, animate and inanimate objects, and light, were found, unfinished, among his papers. He also wrote volumes of essays on a wide range of subjects, reports on the legal cases he was involved in, advice on the union of the crowns of England and Scotland, on travel, the Church of England, and the history of the reign of King Henry VII; he was tackling

Henry VIII when he died. Other works included translations of psalms and prayers, masques and endless letters.

It is no wonder his legacy has morphed to such a degree it obscures him almost completely. Baconism, Baconian science, Baconian method – there are so many Bacons, so many ways he has been interpreted and adopted by people with different priorities. Some people are convinced William Shakespeare was just a pseudonym, and that Bacon was the author of all the plays and sonnets, although when he would have had time to write them is another matter. The Puritans promoted his ideas during the Interregnum, the French based their Académie des Sciences on his philosophy (much to Descartes' annoyance) and, when the monarchy was restored in Britain, he was taken up by the founders of the Royal Society. This would doubtless have pleased him, but in his own lifetime, his scientific works received little attention and no acclaim; he was known as an essayist, a politician, a lawyer, a brilliantly clever man who was neither generally well liked nor afraid to make an enemy. A man before his time.

Francis Bacon is usually considered as a starting point, a spark that ignited the fire that raged across centuries and countries, the 'scientific revolution' – if you subscribe to that reading of history. In this story, we are not endeavouring to look forwards and plot the fault lines of his future influence, rather to look back into the sixteenth century and ask where he got his ideas from. How did he visualise Salomon's House and set it down on the page?

The first thing to appreciate is that Bacon was not a scientist. His contribution lay not in what he discovered, but in the method he proposed for enquiry into the natural world. In this way he differs from the other figures we have met. He did not make astronomical observations, although he did own instruments like astrolabes; he was not even convinced by Copernican theory.

There is no evidence to show that he employed assistants or carried out experiments (apart from the strange and likely apocryphal tale of his death, recounted by John Aubrey, in which he caught a chill after trying to preserve a dead chicken by filling its corpse with ice), or was involved in the empirical study he promoted as the means to understanding and gaining dominion over nature, and improving the lot of humanity in the process.

He was, however, utterly convinced of the importance of technological inventions, in particular the printing press, gunpowder and the magnetic compass, which 'changed the whole face and state of things throughout the world; the first in learning, the second in warfare, the third in navigation; whence have followed innumerable changes; insomuch that no empire, no sect, no star seems to have exerted greater power and influence in human affairs than these changes'.[2] He introduced the idea that science should be dedicated to the improvement of human life, and that it should be funded by the state, a guiding philosophy that has been influential ever since.

Among the towering boxes of documents and books Bacon left behind him, there is surprisingly little about his childhood and adolescence. He was educated at home in the schoolroom next to his brother Anthony, by tutors chosen by his extraordinarily erudite mother, Anne Cooke. Anne was born in 1528 to a respected gentry family in Romford, Essex. She had four sisters, and each of them went on to marry major players on the Elizabethan scene. They were exceptionally well educated – 'all most eminent scholars' – whose parents, Anthony and Anne, had been influenced by humanists like Thomas More and Erasmus, but, unlike More, were also firm believers in the new reformed religion.[3] A visitor to their home in the 1540s described it as being 'like a small university in which women's education flourished'.[4]

But in Tudor England, queens excepted, no amount of erudition would give a woman access to official positions of power and the Cooke sisters had to make their careers behind the scenes, in their husbands' shadows. Anne did, however, enjoy professional success and renown as a translator. In 1548 she published five sermons by the Italian preacher Bernardino Ochino whom Archbishop Cranmer had invited to England to help establish the Reformation. Henry VIII had died the year before, leaving the way clear for the Protestant circle that surrounded the young king Edward VI and included Anne's father Anthony, who was knighted and appointed a royal tutor. Anne, aged nineteen, was so affected by the power of Ochino's words (which were said to 'make the very stones weep') she translated them into English and wrote a preface, and they sold well.[5] In 1551, she translated and published fourteen more. These, too, were widely read and she gained a reputation for piety and scholarship.

This was a halcyon period for English reformers, busy establishing a new version of Christianity, full of hope and resolve for a better future. William Cecil was another major figure in this group, and when not tied up with political matters, he was paying court to Anne's older sister, Mildred. The two married and in the meantime, Cecil introduced Anne to a friend from the Inns of Court, the prominent lawyer Nicholas Bacon. Bacon was eighteen years older than Anne, and already had a wife and children. He was entirely self-made, one of many who had ascended the Tudor social ladder in the slipstream created by the likes of the Thomases, Wolsey and Cromwell (sons of a butcher and blacksmith, respectively). His father, a sheep-reeve to the Abbot of Bury St Edmunds, had sent him to the abbey school, from where he had won a scholarship (like so many others we have met on this journey) to university, in this case Corpus Christi College, Cambridge. From here he progressed to Gray's Inn, arriving in 1523 when Thomas Wolsey was residing in pomp at Hampton Court and Thomas Cromwell had just taken up a seat in parliament.

It was a good time to be a clever young man looking to rise in the world and Bacon rose fast. By 1552 he was an MP and Treasurer of Gray's Inn, a man of means and talent. He used his various positions to amass wealth and property, ballast against the vagaries of misfortune that could easily beset someone who had scaled the heights of Tudor power – there was a long way to fall. Bacon's job at the Court of Augmentations, which had originally been set up by Cromwell, involved overseeing the sale of confiscated monastic properties. This put him in the position to acquire several manors and he gradually built up a portfolio of property. For a self-made man this was essential. Tudor society was entirely structured around inherited wealth and power, and those born into the lower echelons had to create a foundation of wealth on which to build their careers and provide for their heirs, carefully husband their resources, and invest wisely to be able to keep their place on the political stage. The fruits of patronage were far from sufficient, if they materialised at all.

In December 1552, Bacon's wife Jane died suddenly, leaving him with six children under twelve. He waited just a few weeks before he and Anne were married – either he was smitten or desperate for a woman to take over his household. Either way, they appear to have had a happy marriage and the Cooke family was a definite step up for him as a simple yeoman. The reign of Edward VI was a brief period of opportunity for English reformers, but when he died in 1553, his sister Mary returned to England to the embrace of the papacy, leaving families like the Bacons and the Cecils vulnerable. Seeing the danger, Anne lost no time in rushing to the new queen's side at Kenninghall in Norfolk, near to their home at Redgrave, and declaring allegiance to her. Mary appointed her as a lady-in-waiting and for the next six years Anne served her Catholic mistress with devotion, while quietly remaining true to her own faith.

Anne's quick thinking paved the way for other prominent Protestants to serve the new queen, including her brother-in-law, William Cecil, who was sent as the Privy Council's representative. Anne had made sure that the queen now 'thought verye well of her brother Cicell'.[6] He was pardoned and allowed to continue as an MP, while losing the more important positions. Other Protestants fled to safe places on the Continent to wait out her reign, and many reconverted to Catholicism.

In the early years of their marriage, Anne gave birth to two daughters who both died, as so many babies did in this period. Nicholas Bacon wrote a poem to his wife which gives a moving insight into their relationship:

> Calling to mind my wife most dear
> How oft you have in sorrows sad
> With words full wise and fully of cheer
> My drooping looks turned into glad

He goes on,

> Thinking also with how good will
> The idle times which irksome be
> You have made short through your good skill
> In reading pleasant things to me.
> Whereof profit we both did see,
> As witness can if they could speak
> Both your Tully and my Seneke [Cicero and Seneca].[7]

Their intellectual bond was clear – Nicholas speaks of Anne as his equal, valued, respected and loved – not something many wives in this period could expect from their husbands.

Queen Mary died in November 1558, childless and alone, on the very same day that her staunchest supporter Cardinal Pole also passed away. These two deaths spelled the end of Catholicism as the official religion of England. On Elizabeth I's accession to the throne, she appointed William Cecil as secretary of state and Nicholas Bacon as lord keeper of the Great Seal, knighting them both at the same time. Redgrave was too far from London for someone who needed to be in constant reach of the court, so Nicholas bought a property at Gorhambury near St Albans that had been part of the abbot's estate.

Hertfordshire, with its proximity to the capital, rang with construction work in this period as thirty large new houses, most of them former monasteries, were created. Rebuilding at Gorhambury began in 1563. The old medieval hall was demolished, making way for an unusual white rendered house, with modern touches imported from the Continent – classical details, marble and portraits – similar but on a more modest scale to the Cecils' abode Theobalds, just a couple of hour's ride southeast. Bacon, Anne and their two young sons (Anthony was born in 1558, Francis in 1561) were soon settled in their elegant, sophisticated new home, on which Bacon had spent over £3,000, designed to reflect the ruins of the Roman town of Verulamium and allow the family to indulge the ideal of 'otium' – free time – spent in beautiful surroundings like the classical poets Anne and Nicholas so admired. They enjoyed modern luxuries like water piped from the nearby river which fed the house and the many water features in the gardens, and the spacious rooms were filled with books, portraits and lovely things.

Upstairs in the airy school room, two small boys sat at their desks, books open and pens at the ready. Francis and his older brother Anthony were educated at home by tutors, learning Latin fluently by reading the favourites of the Elizabethan schoolroom – Ovid, Horace, Cicero and Virgil – alongside the Bible as devout Protestants. Given their parents' adherence to the new religion, it is probable that Melanchthon played a role, either via his own writings or his recommendations for education.

Both boys showed an early aptitude for learning, but Francis was especially precocious. So much so, that when the time came for Anthony to leave for university, Francis (although just twelve) went along too. They set off to Cambridge in April 1573, just a few hours' ride from Gorhambury. They enrolled as students and moved into rooms together at Trinity College, founded just a few decades before by Henry VIII and coincidentally the place where Dee had worked for a while as Under-Reader in Greek. John Whitgift, the future Archbishop of Canterbury, was master of the college and personally responsible for the young Bacons. They lived in his quarters, and he supervised all aspects of their university lives, even ordering special meals of boiled mutton for them to improve their health. He bought books including the *History of Rome* by Livy and Homer's *Iliad* for them to study, furnished their rooms with wall-hangings, maps, gave them the luxury of glazing in their windows, and 'endless pairs of shoes and slippers', every accoutrement suitable for young gentlemen. This set Anthony and Francis on the path to join the Elizabethan elite, giving them a taste for the high life that they never lost and could seldom afford.

Unlike their aristocratic peers, the young Bacons could not finish their education with a grand tour of the sights of Europe. They did both go abroad when they left Cambridge, but in a working, rather than a leisure, capacity. Francis set off for France in

1576 to continue his training as a lawyer as part of the ambassador Sir Amias Paulet's household. He studied civil law and carried out a range of diplomatic duties for the cash-strapped and beleaguered Paulet, whose position as a devout Protestant at the Catholic French court was tricky, to say the least. Just four years earlier, the streets of Paris had flowed with Huguenot blood during the St Bartholomew's Day Massacre, and relations between the French and English crowns were strained, requiring Paulet to walk a diplomatic tightrope. They got used to being given the worst quarters when the court went on progress around the French countryside.

In spite of these pressures, Francis enjoyed his time with the embassy, making life-long friends with one of Caesar Adelmare's sons who was also studying civil law (Caesar was Mary I's physician, originally from Padua) and with the miniaturist Nicholas Hilliard, who painted an exquisite portrait of Bacon with the legend 'If the face as painted is deemed worthy, yet I prefer the mind, in his eighteenth year.'[8] Paulet's wife Margaret was a kind and loving woman who looked after the young men in her husband's entourage, along with her own six children. Francis was especially grateful to her for curing his warts with a piece of lard, not something it's easy to imagine the fearsome Anne Bacon doing – she would have been more likely to recommend a rigorous schedule of prayer as a remedy.

During his time in France, Bacon first became acquainted with the money problems that would beset him for the rest of his life. Short of cash to fund his lifestyle at the notoriously ruinous French court, he turned to his friend Thomas Bodley, himself a supreme 'intelligencer', to ask if he could use his position in France to send useful information back home. Bodley sent him an advance of £30 and instructions on how to become an elite spy: to provide.

knowledge of the country and the people among whom you
live…enquire carefully, and further help yourself with books that
are written of the cosmography of those parts, you shall sufficiently gather the strength, riches, traffic, havens shipping,
commodities…note their buildings, furnitures, their entertainments; all their husbandry, and ingenious inventions.[9]

At this moment of heightened religious tension, any kind of information could be relevant but Bacon was expected to use his position and, even more so, his intellect, to discover the peculiarities
of the French state, its laws, its plans, its revenues and its
weaknesses.

It is not surprising to hear the future founder of the Bodleian
Library recommending the perusal of French books; he was
already a committed collector and his father had set up a press in
London. The Bodleys were one of the Protestant families who had
spent Mary's reign in the safety of Europe, in their case Switzerland,
hotbed of Calvinism and the new educational ideas Thomas experienced as a student in Geneva. Bacon echoed Bodley's wise words
years later in *The New Atlantis* when he described how the inhabitants of Bensalem sent two ships out into the world every twelve
years to discover 'knowledge of the affairs and state of those countries…and especially of the sciences, arts, manufacturers, and
inventions…and withal to bring unto us books, instruments, and
patterns in every kind'.[10]

In January 1578, Paulet wrote to Nicholas Bacon to let him
know that Francis had moved out of the embassy quarters into the
household of a civil lawyer, so he could see law being practised
and improve his French. At around this time Francis asked Paulet
to write to his father about 'his intended voyage into Italy'. Like so
many of the young men in this story, he yearned to see the canals

of Venice, the art of Florence and the lecture halls of Padua. His half-brother, Edward, several years older, had spent the previous couple of years travelling around the major centres of Protestantism, meeting the great and the good, carrying letters from Hubert Languet in Strasbourg to Joachim Camerarius the Younger, busy at the imperial gardens at Vienna. Edward had even spent a week in Venice and two at Padua before travelling to Paris via Geneva and Zurich, where he doubtless regaled his little half-brother with wild tales of running the gauntlet of 'all kind of vice'.

Much to his chagrin, these dangers were considered to be too much for Francis. On Paulet's unequivocal advice that 'no English gentleman now being on this side the Seas should live in greater danger in Italy than your Lordship's son', Nicholas Bacon decided that his plan was out of the question. The name of Bacon was too well known for its radical Protestantism to risk travel into Italy and the jaws of the inquisition.[11] This was partly thanks to his mother's publications and resulting reputation. The frustration of being the youngest child of such notorious parents must have been hard to bear; this was as close as Francis ever got to the wonders of Italy. The following year his father died and Francis returned to England, never to cross the channel again.

Nicholas Bacon had worked assiduously to ensure that each of his children, and in particular his five sons, were well provided for so that the name of Bacon, so recently risen from the crowded fields of the yeomanry, would remain among the distinguished ranks of gentlemen from now on. When he died, he had made provision for his four eldest sons, but not, as yet, for Francis, leaving him with almost no inheritance to speak of. Gorhambury was to go to Anne for her lifetime, and after that to Anthony.

This left Francis with the inclinations of a nobleman and the income of someone much further down the social scale. He had to earn money, so he threw himself into a career at the Bar and moved into rooms in Gray's Inn. While the older boys squabbled with their stepmother over the will, and Edward and Anthony set off for the Continent again, Francis progressed quickly up the legal ranks and wrote regular letters to his uncle and aunt, the Cecils, asking them to put in a good word for him with the queen. Nothing came of this; Cecil seemed to suggest that so long as they were healthy, Francis and Anthony 'wanted nothinge'. He also wrote to his sister-in-law Anne that 'I am of lesse power to doe my frends good than the worlde thinketh,' something Dee could painfully attest.[12] If Elizabeth I was the empress of the empty promise, Cecil certainly wasn't much better. Bacon's chaplain and biographer William Rawley put it elegantly when he wrote, 'Though she [Elizabeth I] cheered him much with the bounty of her countenance, yet she never cheered him with the bounty of her hand.'[13] It is true that the government was always short of money, but equally true that they could summon up funds when they felt like it, in the case of the alchemist de Lannoy for instance.

Francis did manage to get a seat in parliament, thanks to connections through his half-sister and his godfather the Earl of Bedford, and took his place for the first time in early 1581. Seats in various constituencies followed, and Bacon became a familiar figure in the House of Commons, known for his eloquence and somewhat disagreeable character. He did not make a good start, unwisely opposing Elizabeth during a debate on public fasting, damaging his prospects of preferment even further. Francis had uneasy relationships with many powerful people, including his cousin Robert Burghley and his ultimate nemesis, Sir Edward Coke, who devoted almost as much time to undermining Bacon as he did pursuing his own career as a lawyer and statesman. His fortunes began to transform after James I arrived from Edinburgh to take up the throne, although nothing happened overnight. He was knighted among 300 others as part of the coronation but spent a good part of 1603 hoping for an official position that did not materialise, so he turned to his philosophical work instead.

In 1604 Francis was formerly appointed as a 'Learned Counsel'; over time, the new king 'raised and advanced him nine times', making him attorney-general, lord keeper of the Great Seal and the first-ever lord chancellor. His knighthood was elevated to a baronetcy and, in 1621, he was dubbed Viscount St Alban. It must have felt as if the sun had finally come out from behind a cloud to shine on his years of hard work. Money was still an issue, and always would be – the learned counsel position only paid a modest £60 a year – but at last he had started living up to his father's shining example and could feel proud of his position in society.

All this time he had been working as a lawyer and sitting as an MP, Bacon had been pursuing other interests in his own time. In 1605 he published *The Advancement of Learning*, the first step in revealing his plan to comprehensively overhaul the intellectual world, addressed directly to King James I. This energetic work in two books covered every aspect of learning, its defence, transmission and practice from scholasticism to the cosmetic arts, eloquently demonstrating the extent of Bacon's ambition, and the reach of his intelligence. It explains the importance of practical and experimental research using up-to-date technology, alongside traditional reading, as a means of acquiring knowledge, 'For we see spheres, globes, astrolabes, maps, and the like, have been provided as appurtenances to astronomy and cosmography, as well as books.' He recommends planting gardens for growing simples to make medicines, and using 'dead bodies for anatomies', as Vesalius had done in the anatomy theatre at Padua.[14]

As we know, Francis had never been to Padua, or any other European centre of learning bar Paris, so where did he get all his information about how one should be designed? *The Advancement of Learning* was published in 1605 but Bacon had been thinking about the subject for several years. In 1592 he had written to his aged uncle William Cecil, Lord Burghley, 'I have taken all knowledge to be my province; and if I could purge it…I hope I should bring in industrious observations, grounded conclusions, and profitable inventions and discoveries; the best state of that province.'[15] Two years later he wrote the *Gesta Grayorum*, an entertainment to be performed at the Inns of Court in front of Elizabeth in which he praised the achievements of the learned rulers of the past and listed the four things a prince should possess in order to promote learning:

The collecting of a most perfect and general Library wherein whatsoever the Wit of Man hath heretofore committed to Books of worth, be they ancient or modern, printed or Manuscript European, or of the other Parts, of one or other Language…

Next, a spacious, wonderful Garden, wherein whatsoever Plant, the Sun of divers Climates, out of the Earth of divers Moulds, either wild, or by the Culture of Man…this Garden to be built about with Rooms, to stable in all rare Beasts, and to cage in all rare Birds; with two lakes adjoining, the one of fresh Water, and the other of salt, for like variety of fishes: and so you may have in a small compass, a Model of Universal Nature made private.

The third, A goodly huge Cabinet, wherein whatsoever the Hand of Man, by exquisite Art or Engine, hath made rare in Stuff, Form, or Motion, whatsoever Singularity, Chance and the Shuffle of things hath produced, whatsoever Nature hath wrought in things that want Life, and may be kept, shall be sorted and included.

...fourth, Such a Still-house so furnished with Mills, Instruments, Furnaces and Vessels, as may be a Palace fit for a Philosopher's Stone.[16]

Here is the template for what would, in later years, expand into Salomon's House. In England, places with a garden, a library, a cabinet of curiosities and a laboratory were few and far between, but we are already familiar with one such location, and it is highly likely that Bacon was too. On 11 August 1582 Dee records a visit from a 'Mr Bacon'. There were several 'Mr Bacons' in Francis's immediate family, never mind in the country at large, but we can be sure that this was Francis because he arrived with 'Mr Phillips of the custom house', a close friend whose son Thomas, a master cryptographer, had been part of Paulet's entourage in France.[17]

The house at Mortlake was certainly the most important centre of knowledge in those decades, in terms of its use and the breadth of Dee's library. He also had stills and laboratories with specialised equipment, and owned several items that would be considered curiosities – although these were not displayed in priceless inlaid cabinets as collections on the Continent were. As far as the garden was concerned, we know that Dee 'hired the barber of Chiswick, Walter Hooper, to keep my hedges and knots in as good order as he had them then'.[18] He did not have the means for anything like the lakes and proto-safari park Bacon describes, but the gardens at Gorhambury would have been a closer model and Francis' interest in botany flourished after he took over the house on the death of his mother in 1610.

Francis also appears to have been influenced by Dee's *Mathematical Preface*, written as an introduction to the first English translation of Euclid's *Elements*, the foundation text on geometry. Published in 1571, it was not the success Dee and Henry

Billingsley, the translator, had hoped for and was not reprinted until 1651. Mathematics was still a marginal subject in England, viewed with suspicion, and this was only encouraged by Roger Ascham in his bestseller, *The Scholemaster*, published the year before. Whether or not he had seen the *Elements*, Bacon echoed Dee's pleas for mathematics to be put into the practical service of humanity, to be learned and used by surveyors, sailors and builders, to become part of the fabric of everyday life.

Dee and Bacon were before their time, judging by how the books were received. When copies of *The Advancement of Learning* arrived from the printer, Henry Tomes at Gray's Inn Gate, Francis eagerly packaged them up and sent them to influential friends and acquaintances, including his cousin Robert Burghley, Thomas Bodley in Oxford and the Earl of Northampton, whom he asked to present a copy to the king. Unfortunately, it was late October 1605. A few days later a plot to blow up the Houses of Parliament was discovered, and in the ensuing furore, no one took much notice of the book.

Francis also sent a copy to his friend Tobie Matthew, who was travelling in Italy. Theirs was a lifelong friendship that survived Tobie's conversion to Catholicism and long absences when he was living on the Continent. They corresponded regularly, sending each other books and ideas, and Tobie kept Bacon up to speed with what was going on in Europe. Tobie spent time in Rome around this time and may well have told his friend about the recently founded Accademia dei Lincei for the exploration of the mathematical and physical sciences. *The Advancement of Learning* was written in English; if it was to make any impact abroad it would have to be translated into Latin, something that did not happen until 1622. Tobie translated several of Bacon's works into Italian.

In 1606 Bacon married Alice Barnham, who he had first met when she was a child of eleven. She was now a young teenager, and he was forty-five, attracted no doubt by the £300 a year that would come to her husband and the further £6,000 she was set to inherit – he desperately needed money, and society expected him to marry. Anne must have rejoiced (in so much as her strict Puritan beliefs allowed) that her son was finally settling down, but the 'childer's childer' she had dearly hoped for never appeared, and there is, in fact, evidence that both Francis and Anthony were homosexual, or bisexual. Anthony never married and was accused of sodomy whilst living in Navarre; relations between the formidable, devout mother and her two somewhat dissolute sons were often strained. The letters that passed between them reveal frustrations on both sides, often caused by money, which none of them had enough of. They wrote to her in Hertfordshire 'requesting carpets and cups, pictures and beds, globes and astrolabes' to be sent up to their rooms at Gray's Inn; meanwhile she struggled to keep the estate going with dwindling resources.[19]

Anne lived to be eighty-two; the last years of her life marred by illness. Her funeral was only attended by Francis and his friend Sir Michael Hicks, a poor showing for someone who played a pivotal role on the political stage supporting her husband and later in the development of the Puritan movement. She is buried somewhere in the churchyard of St Michaels, St Albans, without any kind of memorial – yet another brilliant woman who has faded from history. Her only monuments are the books she published, and a dedication in a volume by the French Calvinist Theodore Beza.

James I does not seem to have shown any particular interest in *The Advancement of Learning*, in spite of his own reputation for it. His interests lay elsewhere, in the dangers of witchcraft, the divine right of kings and spending time with his favourites like George

Villiers. Ironically, he was one of the very few people in England who could have given Bacon a first-hand description of Brahe's palace on Hven because as we saw earlier, he had visited there during his trip to Denmark to pick up his young bride, Anne, sister of Christian IV. This does not appear to have kindled any desire in him to set up something similar in his own realm, but it is possible Bacon discussed it with him, the courtiers who had accompanied him or Anne's Danish attendants – the close connections between the two courts meant there were any number of ways information could have passed between them. Equally, Bacon could have got hold of a copy of Brahe's descriptions of his instruments and observatory on Hven, which had recently been published in Prague, and he would certainly have been aware of Rudolf II's extraordinary court there.

There were plenty of figures coming and going between London and Prague: Sidney and Dee, as we have already seen, but also Roger Cook, the melancholic character who oversaw the stills at Mortlake and possibly helped furnish the laboratories. Having worked for Dee from 1567 to 1581, he went on to build still-houses in the Tower of London for Walter Raleigh and the Earl of Northumberland while they were imprisoned there. As such he was an instrumental figure in the growth of this type of scientific workspace.

Cook also spent time at the court of Rudolf II in Prague, where he put his talents to good use working for Cornelis Drebbel on alchemical experiments. Drebbel is one of those characters who has unaccountably slipped through the cracks of time. One of the most innovative thinkers of the period, he was born in Alkmaar in the Netherlands in 1572, the first Dutch city to repel Habsburg rule. His education at Haarlem Academy was humanist, artistic and technical, allowing his practical and intellectual talents to flourish. He learned engraving and became interested in creating innovative new machines, working with clockmakers and spectacle makers to learn the cutting-edge skills the Low Countries were famous for. In 1598 he was awarded two patents, one for a fountain that spouted water pumped through lead pipes, the other for a perpetual motion clock; four years later he added another, for an innovative chimney system.

Francis Bacon was apparently one of those deputised to welcome Drebbel and his family to England in 1604 and help them get settled in their quarters at Eltham Palace. With his passion for investigating nature and using its forces to cause wonder and create useful machines, Drebbel was an exemplary proponent of the 'arts mathematicall' Dee had described in his *Preface to Euclid*, just as he epitomised the scholar untainted by Aristotelianism (he did not attend university) promoted by Bacon. He was interested in many things: he worked with metals, magnets, dyes, glass; studied the weather and the effects of water on the planet; experimented with air pressure and mirrors; he was an expert glass-blower; invented a lens-grinding machine, a furnace with a thermostat and a self-regulating dampener; engraved maps; advocated starting a lottery; made telescopes, an incubator, torpedoes, a microscope and a camera obscura, 'which produces marvellous results in the shape of pictures reflected in a dark room. It is impossible to express the beauty in words'.[20]

Drebbel wrote several books explaining his ideas, apologising for his lack of knowledge of ancient authors. In this regard, as in his universal, experimental interests, he is a blueprint for Bacon's new philosopher of science, a man whose brain was not filled with woolly Aristotelian theories but actively engaged with many aspects of the natural world. It is a mystery why Drebbel is not a more celebrated figure, especially in Britain. Around 1620, he built a functioning submarine which he piloted under the murky waters of the Thames, to the astonishment of assembled Londoners and the king. This vessel, a modified rowing boat with an enclosed roof, the oars sealed into its sides with leather, was floated out onto the river at Richmond and submerged, reappearing three hours later several miles downstream at Greenwich. When Bacon wrote in the *New Atlantis* a few years later, 'we have ships and boats for going under water', he must have had Drebbel's submarine in mind. This apparent miracle was possible because Drebbel had worked out how to extract the 'quintessence of air' – oxygen – from saltpetre and released it periodically into the submarine to keep himself and the fifteen other men inside alive. This inspired Robert Boyle and Robert Hooke's later work, feeding into mechanistic theories that there is indeed a life-giving spirit in the air we breathe.

James I delighted in courtly entertainments that wove symbolism with technical wizardry and artistic brilliance: Bacon, Drebbel and others like Ben Jonson, Inigo Jones and of course, William Shakespeare, all benefitted from this source of patronage. Drebbel created special effects for masques and plays, ghostly illusions and simulations of thunder and lightning. He explained in a work of 1604 on alchemy that saltpetre could be heated to release gases which produced loud banging noises. He also tutored James' eldest son, Henry Prince of Wales, who was growing into an intelligent, cultivated young man, 'a Maecenas of the arts' whose interests were more varied and artistic, if less erudite, than his father's. (In fact, his passion for the visual arts was inherited from his mother, Anne of Denmark, who was said to enjoy the company of pictures more than people.)

In 1610 Henry was set up in his own household based in the palaces of Richmond and St James's, where he created a brilliant court, commissioning works of art, 'entertaining the best engineers and architects of Christendom', and founding a collection he hoped would eventually rival those of the Medici and Rudolf II.[21] This was not to be. Henry died in 1612, aged eighteen, but he was instrumental in introducing the idea to Britain, the first royal to do so; it would be emulated many times in the future and formed part of the same currents visible in Salomon's House.

While Drebbel was living at Eltham Palace, he made a perpetual motion machine for James I, which appeared to work by magic, but was in fact powered by changes in atmospheric pressure, foreshadowing the barometers developed in later centuries. News of this miraculous machine spread far and wide – in Prague, Kepler suspected Drebbel had managed to create a new 'spirit' or source of energy. The two men met when Drebbel arrived there in the autumn of 1610, bringing a perpetual motion machine with him. The timing was not great, however, for Rudolf died just over a year later and Drebbel returned to England, leaving the alchemical experiments and mining projects he had been involved in unfinished. He was another link in the chain connecting Stuart London with Rudolfine Prague, someone who could have given Bacon a detailed account of the wonders he had seen in Bohemia. In 1619 Kepler dedicated his book *Harmonices Mundi* to James I and his sumptuous presentation copy stamped with gold fleur-de-lis is in University College Library, Oxford.

Apart from the short stint in Prague at Rudolf II's court, Drebbel lived in London and worked for the Stuart monarchs, and he died in 1633 at the family alehouse on a quay in the shadow of the Tower. There is no monument to him, no street named for him, and a search of the Bodleian Library catalogue yields just twelve results, of which three are duplicates. Just six are in English, an astonishingly small number for a man who spent most of his adult life in London, holding the post of royal engineer and building extraordinary machines.

In 1621 Bacon's political career came crashing down. He was accused of bribery, impeached and released from his political duties, leaving him free to spend the last five years of his life writing. During this time, he produced *The New Atlantis*, but it was not published until 1967, several years after his death and like many others, was 'A Worke unfinished'. In the book he describes a utopian society where piety, peace and learning go hand in hand. Founded in 300 BCE, the inhabitants of Bensalem (Hebrew for Jerusalem) have been able to develop without the years of scholasticism Bacon saw as having limited the European mind. Several pages follow, listing the numerous parts of the endeavour, which enable the 'fellows' to study every aspect of the natural world in specially designed, controlled environments. These include: caves, high towers, lakes of fresh and saltwater, streams, wells and fountains, houses where they recreate snow, rain, hail and thunder, chambers of purified air, baths of various mixtures to cure diseases, orchards and gardens for botanical study, parks full of animals, brewhouses, kitchens and bakehouses, pharmacies, workshops for manufacturing materials and dyes, furnaces, perspective houses for experiments on light, collections of minerals and precious stones, sound houses, perfume houses, and engine houses where weapons and all types of vessels are made, workshops for clockmaking and a mathematical house where instruments are wrought, and finally, 'houses of deceits of the senses' for investigation into illusions of all kinds. The whole world is here in microcosm, to be explored, understood and controlled for the greater good. Salomon's House is the culmination of all the places we have explored and so much more. It was a prophetic vision of what science could and would achieve.

Thanks to the rediscovery of a manuscript of the greatest utopian vision of them all, Plato's *Republic*, there was a rich seam of this type of writing in the early modern period. Plato's description of the mythical island of Atlantis in the *Timaeus* is an obvious source for Bacon. With its emphasis on moral rectitude and religion, *The New Atlantis* was also clearly influenced by Thomas More's *Utopia* written a century before, and Tommaso Campanella's *City of the Sun*, written in 1602 but not published until 1623. In the latter work, astrology plays a central role in the citizens' lives; Campanella was a stellar magician whose activities gave the Inquisition cause to imprison him for thirty years.

There are also many similarities between Salomon's House and a book called *Christianopolis, An Ideal State of the Seventeenth Century*, written a few years previously by Johann Valentin Andreae, a leading figure in the German Protestant movement and part of the alchemical, Paracelsian tradition to which Dee, Rudolf II and many others adhered. Andreae studied at the University of Tübingen under Kepler's old teacher Michael Maestlin, who may have put them in touch – he and Kepler corresponded for years, until the older man's death in 1630. They were from the same region, born fifteen years apart but both educated at Tübingen, where Kepler lived at the Stift, a Church-funded hall of residence Andreae later helped to restore during the Thirty Years' War. While Maestlin showed Andreae the wonders of maths and astronomy, his other significant early influence was his mother, Maria Moser, who took sole responsibility for him after his father died. They moved to Tübingen so he could be educated, and she worked as the court apothecary at Württemberg from 1607 to 1617.

Unlike Francis, Andreae's devout Lutheranism did not prevent him from travelling extensively in Italy, Austria and France; he also spent time in Switzerland and visited several cities in his native Germany. He worked in Vaihingen as an assistant minister before settling in Calw, not far from Tübingen. He was part of an influential Protestant movement promoting social reform, education and in particular the sciences as a means of achieving utopian ideals and improving society for everyone. This began in Germany and gradually spread across northern Europe to Britain, via works like Andreae's. There is no specific evidence that Bacon read *Christianopolis*, other than the many striking similarities between it and *The New Atlantis*. Both begin with storms at sea and lost mariners, and both describe mysterious, impressive societies where learning and Christian piety are valued above all else. Andreae's work is neater, more structured, and crucially, complete, in contrast to *The New Atlantis* which is very much part of the overarching *Great Instauration* project and was written in the final years of Bacon's life. *Christianopolis* is divided into short chapters with titles including 'Night Lights', 'the Natural Science Laboratory', 'Mystic Numbers' and 'The Temple'. There is even a section on women, and one third of the students in the institution are girls – like Bacon's maternal grandfather Sir Anthony Cooke and Thomas More, Andreae was a believer in the value of female education.

Astronomy does not play much of a role in Salomon's House, reflecting the reality that scientific investigation had broadened out to such an extent it was now one part of a much larger whole, rather than the leading discipline it had been in the previous century. There were many other areas of research now for rulers to spend money on: mining, military technology, glass production (by the early nineteenth century, Britain was the leading producer) and transport, among others. Bacon mentions that the high towers, constructed on mountain tops for ultimate vantage, are used for 'the view of divers meteors; as winds, rain, snow, hail; and some of the fiery meteors also'. The word meteor referred to any type of atmospheric phenomenon and Bacon seems to consider heavenly bodies as an afterthought; meteorology takes precedence.

In the 'mathematical house' astronomical instruments are 'exquisitely made', but Bacon does not elaborate, in contrast to Andreae, who not only includes chapters on astronomy, which 'only the most noble-minded natures have an inclination towards' and astrology, which he expresses doubts about, but refers the reader specifically to Tycho Brahe's work in the section on mathematical instruments. In *Christianopolis*, 'The whole city is, as it were, one single workshop, but of all different sorts of crafts.'[22] There is a laboratory dedicated to chemical science, 'a place given over to anatomy' and a pharmacy, a natural science laboratory, a 'hall of physics' where 'natural history is here seen painted on the walls in detail and with the greatest skill'.

Salomon's House is less traditionally divided according to the old academic disciplines. Bacon is more interested in processes: putrefaction, weathering, refrigeration, desalination, breeding, cultivating and so on. Many of the various places and houses within the institution are, in fact, located outside, in nature – on mountain tops or deep underground. These bring to mind one of Drebbel's more memorable experiments in 1620 when he tested a salt-based air conditioning system on a hot summer's day in Westminster Great Hall. The results were rather too extreme for the king and his entourage, who rushed out, trembling with cold. Did he and Bacon discuss this kind of thing? They must have known one another, and with their shared interests, it's hard to believe they didn't confer, that Bacon didn't visit him at his laboratory in Eltham Palace.

Because he was not a scientist, Bacon was able to take a step back and really look at how investigation into the natural world was carried out, to assess the approaches, methods and traditions, before proposing a new way forward. Part of this was because he was a lawyer, with a brilliantly logical mind. It was also a question of timing; Bacon realised that the chaos of intellectual possibilities that characterised the late sixteenth century could not continue; it needed to be organised into a system that would rank certain areas over others but, most of all, set down new methods of obtaining knowledge to unlock progress. By looking at these stargazers' palaces we can see the ebb and flow of scientific ideas, the development of scientific practice and the related changes in the epistemological structures in this singular period in the wider history of knowledge. As observation gradually triumphed, scholars found new roles in society and the topography of the world of learning shifted. Bacon's vision for centres of learning was taken up and realised in the following centuries. Many of his ideas have remained essential to scientific culture ever since.

EPILOGUE

BEYOND

If we survey the map of Europe again in the middle decades of the seventeenth century, the north has been transformed. The old Renaissance centres of Italy are still lit up, and there are flourishing universities and courts in Spain. Paris is a growing centre of learning and commerce, but it is overshadowed by its neighbours in Germany and the Low Countries. Although beset with one of the most vicious wars the continent has ever seen, they are home to impressive cities where knowledge and technology are being forged. Amsterdam is radiant, at the height of its Golden Age, the centre of global trade, culture and innovation. Copenhagen and Stockholm are also shining; Sweden's learned queen Christina has claimed what remains of Rudolf's collections in Prague. Over the narrow channel to the west, London crackles with energy. In 1500, its population was 50,000. By 1700, it had increased ten-fold, with half a million souls working to build the greatest city on earth.

Commerce and technology led to wealth-creation and changed the scientific landscape during this period, as did the establishment of Protestantism, which promoted literacy through its emphasis on reading the gospels. Melanchthon's educational programme and scholarships expanded the pool of intellectual talent, making it possible for someone like Ursus, the son of a pig farmer, to educate himself and reach the heights of imperial mathematician – social mobility on a scale that would still impress today. Religion and science were intertwined, not always in agreement, certainly, but impossible to divide. Every person we have encountered lived in a society that was structured around the Church; few would have doubted the existence of God; most would have been devout. Where possible, the Protestant mission pervaded education to establish its dominance. University rectors had to impose Lutheran orthodoxy; students were required to adhere to the Augsburg Confession; curriculums were designed to ensure certain interpretations of the scriptures. The natural world was God's masterpiece, so studying it was a way of getting closer to Him. This was true of both the Catholic and Lutheran Churches (and Islam too). In the seventeenth century, religion and natural knowledge (astronomy, alchemy and so on) were interdependent, using each other to provide mutual legitimacy.

The violent interaction between the competing Christian faiths caused migrations of skilled individuals, among them the Protestants who fled Habsburg persecution in the Low Countries and the Huguenots who left France after the St Bartholomew's Day Massacre. Many travelled to England or Scandinavia, where they built communities of specialists like instrument makers; by the early eighteenth century, London would be the epicentre of this industry, taking over the mantle formerly held by Nuremberg and Louvain. In the late sixteenth century, Prague, although a Catholic city in the Holy Roman Empire, was a magnet for Protestant scholars seeking religious toleration and Rudolf II's generous patronage. There was also persecution within the new faith, however; Luther had hardly put down his hammer after nailing the *Ninety-Five Theses* to the church door before alternative denominations began to spring up around him. By refuting the papacy, he opened the floodgates of doctrinal rebellion through which flooded Calvinism, the Anabaptists, Methodists, Philippists and various others over the following decades. Relations between them were often violent and the oppression of members of a competing sect wasn't unusual – Caspar Peucer spent twelve years in prison because he was suspected of being a Calvinist by the Lutheran authorities.

In general, Protestant societies tended to be less rigidly structured. Women often had more freedom, both as patrons and practitioners, and there was greater social mobility, providing opportunities for culture to flourish. The reformed Church, in its infancy, lacked systems of surveillance like the Inquisition or centuries-old traditions of imposing doctrinal conformity. This gave scholars greater liberty than their southern counterparts. The new religion did occasionally confront scientific ideas – Copernicus' heliocentric universe was criticised and refuted by Protestants in the sixteenth century but they did not ban its teaching for almost two centuries as the Catholic Church did. This, and the Protestant emphasis on industry and innovation, helped foster an environment in which northern Europe could flourish intellectually and commercially.

People began investigating nature in a multitude of different places. Hundreds of new universities were founded, many as part of the establishment of Protestantism as a major religion. Medical students thronged anatomy theatres to watch dissections, rare plants blossomed in botanical gardens, cabinets of curiosity filled with wonders and libraries proliferated. In 1669, Leiden became the first university to have a laboratory. Thousands of workshops produced an array of gadgets and instruments that increased every year. If you wanted to buy a quadrant, you could go to a shop to choose one, provided you had the money. You might have even owned a clock. Knowledge was becoming more accessible all the time. Many people still carried out scientific work at home but there was a new sense of community between them.

None of the centres of learning we have visited survived much beyond their founder's lifetime. Almost all of them ended their lives condemned, in poverty or in exile. Brahe had to leave Denmark; Dee died in penury; Mercator never returned to Louvain; Rudolf was deposed. Their laboratories, collections, libraries and observatories were destroyed and dispersed. Wilhelm was respected and revered after his death, but his son Moritz was not interested in continuing his astronomical legacy, focusing on alchemy instead. From the 1660s onwards, science began to have permanent homes in the major cities of northern Europe, places where generations of scholars could work together. In Paris, the Académie Royale des Sciences was founded and in London, the Royal Society. This made certain things possible for the first time: knowledge could be safely stored and assessed; ideas could be shared easily, as could equipment; long-term, large-scale projects became practicable; and scientific investigation became part of the establishment. However, these changes did not take place overnight: it took centuries for them to be realised.

There had been similar institutions before which serve as precedents. In the early sixteenth century the Spanish and Portuguese monarchs founded institutions dedicated to trade, which encompassed teaching and research geared at improving navigation, creating accurate maps and making instruments. The Academia dei Lincei was founded in Rome in 1603, boasting Galileo Galilei and Giambattista della Porta as members. But it closed its doors after just a few years. In 1657, Leopoldo de' Medici established an

Academy of Experiment in Florence, playing a personal role in its management, while in Germany, the Collegium Naturae Curiosorum (later known as the Leopoldina) was founded in Schweinfurt in 1652 as a centre of medical research that soon expanded to include all types of experimentation, funded by the state. In London, Gresham College was founded by the merchant philanthropist Sir Thomas Gresham in 1597 to provide free lectures on astronomy, geometry and rhetoric for anyone who was interested. Gresham, who had no children, left a generous endowment to fund the seven professorships. Sir Christopher Wren was professor of astronomy there in the mid-seventeenth century; he and a group of his friends would gather in his room after lectures to discuss their interests and experiments – this group went on to found the Royal Society in Gresham College.

The new academies were still strongly influenced by court structures and traditions. Both the Royal Society and the Académie Royale des Sciences had royal patronage, but Bacon's plea for comprehensive state funding was not realised at the Royal Society for a long time. Initially, it functioned as a kind of gentleman's club and was 'relatively impoverished even with its royal charter of 1662'.[1] It was a different story at the Académie in Paris. The state provided funding each year and guidance on avenues of research. This formalisation gave members confidence, authority and access to an established network and shared bodies of knowledge. The funding ensured longevity, a crucial change from the centres of learning in the previous century. Their status rose and became official, part of the general trend we have traced, beginning with Dürer, regarding artists, instrument makers and the multitude of practitioners we have encountered. The explosion of material culture and technology helped affect this change, facilitated by patrons like Wilhelm IV and Rudolf II who valued and shared in their work as never before.

The royal charter may not have brought funding, but it did give the Royal Society the power to publish whatever they wished, without having to submit to the censors, and this was priceless. It meant information could be printed quickly, and it provided authority. In 1665, Henry Oldenburg established, edited and paid for the Society's publication, the *Philosophical Transactions* (the oldest continuously published scientific journal), which contained all the data, meeting minutes and records, and disseminated them to the world so that new theories could be discussed, tested and approved. It serves as the origin of the peer review system that underpins academia today. Correspondence networks grew denser and bigger, now stretching around the globe as imperial expansion took off and letters and reports from the four corners of the earth were written up and published in journals.

During the sixteenth century, new roles emerged in the scientific enterprise and specialisation increased. In astronomy this was epitomised by Brahe's skill as an observer, Wittich's mathematical advances and Kepler's talent for visionary thought. Bacon built on this when he allocated the fellows in Salomon's House. Divided into teams of three, they focus on separate tasks – collecting information, carrying out experiments, writing and processing results, directing new areas of research – each group working in synchronicity with the whole. The image of the Renaissance polymath in a study that is at once a library, an observatory, a museum and a laboratory was starting to fade.

Many still believed in astrology, but it was losing its intellectual integrity and central position on the map of knowledge, slipping out towards the mystical fringes where it resides today. There was still a wide diversity of opinion and belief, and much contradiction and confusion, which became problematic as other disciplines developed formal, commonly accepted bodies of fact.

Serious scholars and sceptics throughout the sixteenth and seventeenth centuries highlighted the need to:

> refute all the wickedness and wantonness of those who captivate
> the mind of ignorant people with a mass of lies partly by their
> viciousness and partly by abusing the gullibility of the multitude
> to earn unjust and fraudulent profit. These people have always
> been buzzing about the marketplaces in great numbers, and
> nowadays there are even more of them.[2]

Increasingly, scientific practitioners (they cannot be described as
'scientists' quite yet) focused on prediction in other areas – meteorology for example, which was closely linked to astrology and
astronomy. Brahe was a leader in this field, gathering weather data
from different locations to compare them, and Dee's diaries are
full of similar notes. Of the many aspects of science that involve
foretelling the future, meteorology is perhaps the most prevalent
in the modern world.

While astrology diminished, her traditional 'handmaiden',
astronomy, became more scientific and technological, but was also
marginalised, to an extent, by the multitude of new disciplines.
Once rulers no longer relied on regular astrological predictions,
astronomers could focus instead on the serious business of the
cosmos, and they were given official positions as leaders of state-
funded observatories. In 1675 John Flamsteed was appointed the
first astronomer royal, installed in a brand-new observatory at
Greenwich in south-east London. This was a seminal moment for
astronomy in the British Isles, but old habits die hard – Flamsteed
marked the occasion by casting a horoscope for the new foundation. Despite writing critiques of astrology, both he and Edmund
Halley provided data for popular almanacs published in the 1690s,

a best-selling genre that remained rooted in the tradition for decades to come. The work of the astronomers we have encountered was used and republished in various guises over the following centuries. For instance Flamsteed's *Historia Coelestis Britannica*, published posthumously in 1725, included Wilhelm IV's observations. He had already referred to the landgrave's precise use of time in a lecture he gave at Gresham College in 1682, bolstering the suggestion that Wilhelm IV and Rothmann's observations, made using Bürgi's instruments and calculations, constituted the most accurate data of sixteenth century astronomy. Brahe would doubtless have had something to say about that.

The development of specialised scientific places like observatories and laboratories has defined our ability to make discoveries and increase our power over the natural world, but there is also a downside. Today, interdisciplinarity in academic knowledge is unusual. In science it is rare, and even within individual disciplines like chemistry or physics, there isn't enough communication or cooperation among practitioners. Conferences tend to attract specialists from the same corner of academia; university structures are (literally and metaphorically) designed to be separate from one another; most undergraduate degrees are very narrow in scope. This makes it easy to get stuck in particular methodologies, even though the value of cross-discipline collaboration is widely recognised. There are signs of change, however. At the Francis Crick Institute in London, 1,500 scientists work in seven 'interest groups' rather than 'traditional divisions or faculties'; this maximises the opportunities for collaboration and innovation. Founded in 2016, it is the largest biomedical facility in Europe, incorporating six major institutions within its shiny new walls. The four hundred or so applicants for each research position advertised show how popular this approach is; time will tell how widespread it becomes.

When it comes to the transformation of scientific culture over the past 500 years, nothing comes close to the role played by technology. In astronomy, the leap that changed everything was the telescope. It heralded a new era, ending millennia of 'naked eye' astronomy and providing ground-breaking new insights into our universe. The work of mapping the stars continues today, using machines beyond even Kepler's wildest visions. On Kitt Peak in southern Arizona, the Dark Energy Spectroscopic Instrument uses 5,000 'eyes', moving robots which observe hundreds of galaxies simultaneously, looking deep into time and space to produce a map of the cosmos. As Hermes Trismegistus predicted, magic, or as it's known today, technology, has made us into 'little gods' with untold power over the natural world. So much so that in the past few decades an epoch-defining change has taken place. Francis Bacon's aim was dominion over nature, something we have achieved to our benefit, but at huge cost to the planet. We have realised that nature is finite and incredibly precious, something we need to protect from ourselves. So far, technology has been used to exploit; now we need to use it to save the world, and ourselves.

There is still so much to discover, so much we do not know. And, as with the beautiful, myriad complexities of our planet, the deeper we look into the universe, the more there is to see.

ACKNOWLEDGEMENTS

This has been a difficult book to write, mainly because the sixteenth century was such a vibrant, kaleidoscopic period – there were difficult choices about what to include and what to leave out. As such, it provides one view of knowledge in this period, while there are of course many others. Any omissions or mistakes are entirely my own.

I have been very fortunate to have the support of my agent, Sarah Chalfant, and her team, Alba Ziegler-Bailey, Emma Smith, Samuel Sheldon and Lily Middlemass – their warm encouragement and calm professionalism gave me a firm foundation as I traversed the quicksands of the early modern intellectual realm. My editor Cecilia Stein at Oneworld has been an invaluable source of wisdom and positivity; being able to talk everything through with her has been a vital and enjoyable part of the process. My thanks also goes to all her brilliant colleagues who are involved in the process of getting this book out into the world, and to David Eldridge for designing the most beautiful cover I could have imagined.

I owe thanks to many others who have helped me to write *Inside the Stargazer's Palace*. Bruce T. Moran, Karsten Gaulke, John L. Heilbron, Dione Verulam and Helen Bishop at the

St Albans and Hertfordshire Architectural and Archaeological Society. To Caspar Roscoe for his technological assistance, Antonia and Eugen Nutting for their translations and help trying to track down information on Sabina of Hesse. To Dr Wolfgang Gustav Metzger, Dr Lucy Wooding, Philip Russell, Alasdair Watson, Emilie Savage-Smith and the staff at the Bodleian and Warburg Libraries. To Bert Manton for showing me around his magical workshop.

I also owe a huge debt of gratitude to all my family and friends for their patience, support and encouragement. In particular to Dottie, Sacha, Ellie, Aimster, Sally and Anna, but most of all to my parents, my girls and Mikkel.

This book is dedicated to my brother-in-law Rasmus Bysted Møller, who died in October, the day before his fifty-first birthday. He showed incredible bravery during his long ordeal at the hands of motor neuron disease. He was one of the brightest, most engaging people I have ever met and our conversations were a major source of inspiration for this book.

He is missed every single day.

SELECT BIBLIOGRAPHY

Primary Sources

Augustine, *The City of God* (*De Civitate Dei*), eds. Babcock & Boniface (New York: New City Press, 2012).

Bacon, Francis, *The Major Works* (Oxford: Oxford University Press, 1996).

Borchert, Till-Holger and Joshua P. Waterman, *The Book of Miracles* (Cologne: Taschen, 2013).

Brahe, Tycho, *His Astronomicall Coniectur of the New and Much Admired * which Appered in the Year 1572*, London 1632 (New York: Da Capo Press, 1969).

Brahe, Tycho, *Instruments of the Renewed Astronomy*, eds. Hadravová-Dohnalová, Hadrava, Shackelford (Prague: KLP, 1996).

Cope, Jackson I. and Harold Whitmore Jones (eds), *Sprat's History of the Royal Society* (London: Routledge & Kegan Paul Ltd, 1959).

Crossley, James (ed.), *Dr John Dee: Autobiographical Tracts* (Literary Licensing LLC, 2014). Reprint of original publication in *Remains Historical & Literary Connected with the Palatine Counties of Lancaster and Chester*, Vol. xxiv (London: Chetham Society Publications, 1851).

Dee, John, Bodleian Library MSS Ashmole 487 & 488.

——— Bodleian Library, Ashmole MS 423, f. 294 lost diary.

Dreyer, J. L. E., *Tychonis Brahe Dani Opera Omnia* (Hauniae: in Libraria Gylendalia, 1913–1929).

Fenton, Edward (ed.), *The Diaries of John Dee* (Oxford: Day Books, 1998).

Johnston, Arthur (ed.), *Francis Bacon, The Advancement of Learning and New Atlantis* (Oxford: Clarendon Press, 1974).

Klein, Kevin (ed.), *The Complete Mystical Records of Dr. John Dee, Transcribed from the 16th Century Manuscripts Documenting Dee's Conversations with Angels*, Vol. I & II. (Minnesota: Llewellyn Publications Woodbury, 2017).

Melanchthon, Philip, *Commentary on the Colossians* (Wittenberg: 1529).

Moryson, Fynes, *An Itinerary* (London: J. Beale, 1617).

Osley, A. S., *Mercator: A monograph on the lettering of maps etc.* Also facsimile of 'M.C.'s treatise on italic hand, *Literarum Latinarum.* Translation of Ghim's *Vita Mercatorius* (London: Faber, 1969).

Roberts, Julian, and Andrew Watson, *John Dee's Library Catalogue* (London: Bibliographical Society, 1990).

Sachs, Hans, and Jost Amman, *The Book of Trades* (New York: Dover Publications, 1973).

Shakespeare, William, *Complete Works of William Shakespeare*, eds. Peter Alexander, Germaine Greer and Anthony Burgess (Glasgow: Collins Classics, 1994).

——— *Troilus and Cressida*, David M. Bevington (ed.). Revised edition. (London: Bloomsbury Arden Shakespeare, 2015).

Singer, Samuel Weller (ed.), *The Life of Sir Thomas More by William Roper* (Chiswick: C. Whittingham, 1822).

Spedding, James (ed.), *The Life and Letters of Sir Francis Bacon* (London: Longman, Green, Longman, and Roberts, 1861–1874).

Vesalius, Andreas, *De humani corporis fabrica* (Basel: Joannes Oporinus, 1543).

Vickers, Brian (ed.), *Francis Bacon, The Major Works* (Oxford: Oxford University Press, 1996).

∗
∗   ∗
∗

Secondary Sources

The University of Louvain, 1425–1975 (Leuven: Leuven University Press, 1976).

Catalogue Raisonne of Scientific Instruments from the Louvain School, 1530–1600 (Turnhout: Brepols, 2002).

Andreä, Johann Valentin, *Christianopolis: An Ideal State of the Seventeenth Century*, trans. Felix Emil Held (New York: Oxford University Press, 1916).

Allen, Don Cameron, *The Star-Crossed Renaissance, The Quarrel About Astrology and its Influence in England* (New York: Octagon Books, 1973).

Allen, Gemma, *The Cooke Sisters: Education, Piety and Politics in Early Modern England* (Manchester: Manchester University Press, 2013).

Barnard, John, D. F. McKenzie and Maureen Bell (eds.), *The Cambridge History of the Book in Britain, Vol. IV* (Cambridge: Cambridge University Press, 2002).

Bažant, Jan, Nina Bažantová and Frances Starn (eds.), *The Czech Reader: History, Culture, Politics* (Durham, NC: Duke University Press, 2010).

Bertoloni Meli, Domenico, *Thinking with Objects: The Transformation of Mechanics in the Seventeenth Century* (Maryland: Johns Hopkins University Press, 2006).

Brown, Neil, Silke Ackermann and Feza Günergun (eds.), *Scientific Instruments between East and West* (Leiden: Brill, 2019).

Christianson, John Robert, *On Tycho's Island: Tycho Brahe, Science and Culture in the Sixteenth Century* (Cambridge: Cambridge University Press, 2002).

——— *Tycho Brahe and the Measure of the Heavens* (London: Reaktion Books Ltd, 2020).

van Cleempoel, Koenraad, *Scientific Instruments in the Sixteenth Century, The Court of Spain and the School of Louvain* (Madrid: Fundacion Carlos Amberes, 1998).

Clulee, Nicholas H., *John Dee's Natural Philosophy, Between Science and Religion* (London: Routledge, 1988).

Connor, James A., *Kepler's Witch: An Astronomer's Discovery of Cosmic Order Amid Religious War, Political Intrigue and the Heresy Trial of his Mother* (London: HarperOne, 2004).

Crane, Nicholas, *Mercator: The Man who Mapped the Planet* (London: Weidenfeld & Nicholson, 2002).

Curry, Patrick (ed.), *Astrology, Science and Society, Historical Essays* (Bury St Edmunds: The Boydell Press, 1987).

Daston, Lorraine, and Elizabeth Lunbeck, *Histories of Scientific Observation* (Chicago: University of Chicago Press, 2011).

Dear, Peter, *Revolutionizing the Sciences: European Knowledge and Its Ambitions, 1500–1700* (Princeton: Princeton University Press, 2001).

Drijvers, Jan Willem, and A. A. MacDonald (ed.), *Centres of Learning: Learning and Location in Pre-Modern Europe and the Near East* (Leiden: Brill, 1995).

Eamon, William, *Science and the Secrets of Nature* (Princeton NJ: Princeton University Press, 1994).

Evans, R. J. W., *Rudolf II and His World, A Study in Intellectual History 1576–1612* (Oxford: Clarendon Press, 1984).

Febvre, Lucien, H-J. Martin, *The Coming of the Book: The Impact of Printing 1450–1800* (London: Verso, 1976).

Feingold, Mordechai, *The Mathematicians' Apprenticeship: Science, Universities And Society in England, 1560–1640* (Cambridge: Cambridge University Press, 1984).

Frasca-Spada, M., N. Jardine (eds.), *Books and the Sciences in History* (Cambridge: Cambridge University Press, 2000).

Fučíková, Eliška (ed.), *Rudolf II and Prague: The Court and the City* (London: Thames & Hudson, 1997).

Gaukroger, Stephen, *Francis Bacon and the Transformation of Early-Modern Philosophy* (Cambridge: Cambridge University Press, 2001).

——— *The Emergence of a Scientific Culture and the Shaping of Modernity 1210–1685* (Oxford: Clarendon Press, 2006).

Gillispie, Charles Coulston, and Frederic Lawrence Holmes, *Dictionary of Scientific Biography* (New York: Charles Scribner's Sons, 1981).

Gingerich, Owen, and Robert S. Westman. *The Wittich Connection: Conflict and Priority in Late Sixteenth-Century Cosmology* (Philadelphia: American Philosophical Society, 1988).

Grafton, Anthony, *Defenders of the Text: The Traditions of Scholarship in an Age of Science, 1450–1800* (Cambridge, MA: Harvard University Press, 1994).

Haasbroek, N. D., *Gemma Frisius, Tycho Brahe and Snellius and their Triangulations* (Delft: W. D. Meinema, 1968).

Häberlein, Mark, *The Fuggers of Augsburg: Pursuing Wealth and Honor in Renaissance Germany* (Charlottesville: University of Virginia Press, 2012).

Heal, Bridget, and Ole Peter Grell (eds), *The Impact of the European Reformation: Princes, Clergy, and People* (Aldershot: Ashgate, 2011).

Henry, John, *Knowledge is Power: How Magic, the Government and an Apocalyptic Vision Helped Francis Bacon to Create Modern Science* (London: Icon Books, 2017).

Horwich, Paul (ed.) *World Changes: Thomas Kuhn and the Nature of Science* (Cambridge, MA: MIT Press, 1993). See Swerdlow chapter: 'Science and Humanism in the Renaissance.'

Jardine, Lisa, *Worldly Goods* (London: Macmillan, 1996).

Jardine, Lisa, and Alan Stewart, *Hostage to Fortune: The Troubled Life of Francis Bacon 1561–1626* (London: Orion, 1998).

Karrow, Jr., Robert W., *Mapmakers of the Sixteenth Century and Their Maps, Bio-Bibliographies of the Cartographers of Abraham Ortelius, 1570, Based on Leo Bagrow's A. Ortelii Catalogus Cartographorum* (Winnetka, Ill: Speculum Orbis Press for The Newberry Library, 1993).

Köttelwesch, Sabine, *Geliebte, Gemahlinnen und Mätressen* (Kassel: Hofgeismar, 2004).

van der Krogt, P. C. J., *Globi Neerlandici: The Production of Globes in the Low Countries* (Utrecht: HES, 1993).

Loades, D. M., *Mary Tudor* (Stroud: Amberley, 2011).

Lowry, Martin, *The World of Altus Manutius: Business and Scholarship in Renaissance Venice* (Ithaca, New York: Cornell University Press, 1979).

Markey, Lia (ed.), *Renaissance Invention, Stradanus' Nova Reperta* (Evanston, IL: Northwestern University Press, 2020).

Marshall, Peter, *The Mercurial Emperor: The magic circle of Rudolf II in Renaissance Prague* (London: Pimlico, 2007).

du Maurier, Daphne, *The Winding Stair: Francis Bacon, His Rise and Fall* (London: Virago, 2006).

McCluskey, Stephen C., *Astronomies and Cultures in Early Modern Europe* (New York: Cambridge University Press, 1998).

Metzger, Christof (ed.), *Albrecht Dürer* (London: Prestel Albertina, 2019).

Moran, Bruce, 'Wilhelm IV of Hesse-Kassel: Informal Communication and the Aristocratic Context of Discovery,' in Thomas Nickles (ed.), *Scientific Discoveries: Case Studies* (Dordrecht: Reidel, 1980).

——— (ed.), *Patronage and Institutions: Science, Technology, and Medicine at the European Court, 1500–1750* (Woodbridge: Boydell Press, 1991).

——— *Science at the Court of Hesse-Kassel: Informal Communication, Collaboration and the Role of the Prince-Practitioner in the Sixteenth Century.* Ph.D. thesis (California: University of Los Angeles, 1978).

Mosley, Adam, *Bearing the Heavens: Tycho Brahe and the Astronomical Community of the Late Sixteenth Century* (Cambridge: Cambridge University Press, 2002).

Müller, Jürgen, *The Complete Paintings of Bruegel* (London: Taschen, 2020).

Newman, William R., and Anthony Grafton, *Secrets of Nature: Astrology and Alchemy in Early Modern Europe* (Cambridge, MA & London: MIT Press, 2001).

North, John David, *Cosmos: An Illustrated History of Astronomy and Cosmology* (Chicago: University of Chicago Press, 2008).

Larsen, Carsten (ed.), *Tycho Brahe's Verden, Danmark I Europa 1550–1600* (Copenhagen: Nationalmuseet, 2006).

van Ortroy, Fernand, *Bio-bibliographie de Gemma Frisius, fondateur de l'école belge de géographie, de son fils Corneille et de ses neveux les Arsenius* (Bruxelles: M. Lamertin, 1920).

Osterhage, Wolfgang, *Kepler: The Order of Things* (Cham, Switzerland: Springer Biographies: 2020).

Panofsky, Erwin, *The Life and Art of Albrecht Dürer* (Princeton, NJ: Princeton University Press, 1955).

Park, Katharine, and Lorraine Daston (eds.), *The Cambridge History of Science Vol. 3, Early Modern Science* (Cambridge: Cambridge University Press, 2006).

Parry, Glyn, *The Arch-Conjurer of England, John Dee* (London: Yale University Press, 2011).

Pettegree, Andrew, *Europe in the Sixteenth Century* (Oxford: Blackwell, 2002).

Price, Bronwen (ed.), *Francis Bacon's New Atlantis – New Interdisciplinary Essays* (Manchester: Manchester University Press, 2002).

Pye, Michael, *Antwerp: The Glory Years* (London: Allen Lane, 2021).

Saini, Angela, *Inferior* (London: 4th Estate, 2017).

Sherman, William, *John Dee: The Politics of Reading and Writing in the English Renaissance* (Amherst: University of Massachusetts Press, 1995).

Shumaker, Wayne, and John L. Heilbron (eds.), *John Dee on Astronomy: Propædeumata Aphoristica 1558 and 1568* (Los Angeles: University of California Press, 1978).

Smith, Pamela H., *The Body of the Artisan: Art and Experience in the Scientific Revolution* (London: University of Chicago Press, 2004).

Smith, Pamela H., and Paula Findlen (eds.), *Merchants & Marvels: Commerce, Science, and Art in Early Modern Europe* (New York: Routledge, 2002).

Spencer, Nicholas, *Magisteria: The Entangled Histories of Science and Religion* (London: Oneworld, 2023).

Spring, Deborah, *Mistress of Gorhambury: Lady Anne Bacon, Tudor Courtier and Scholar* (St Albans: St Albans & Hertfordshire Architectural & Archaeological Society, 2021).

Strano, Giorgio (ed.), *European Collections of Scientific Instruments, 1550–1750* (Leiden: Brill, 2009).

Strauss, Gerald, *Nuremberg in the 16th Century* (London: John Wiley and Sons Inc, 1968).

Strong, Roy, *Henry Prince of Wales and England's Lost Renaissance* (London: Pimlico, 2000).

Taylor, E.G.R., *Tudor Geography 1485–1583* (London: Methuen, 1930).

Thomas, Keith, *Man and the Natural World* (London: Allen Lane, 1983).

——— *Religion and the Decline of Magic* (London: Weidenfeld & Nicholson, 1971).

Thoren, Victor, *The Lord of Uraniborg: A Biography of Tycho Brahe* (Cambridge: Cambridge University Press, 1990).

Turner, Gerard L. E., *Scientific Instruments and Experimental Philosophy 1550–1850* (Aldershot: Variorum, 1990).

Risk, R. T., *Erhardt Ratdolt, Master Printer* (Francistown, NH: Typhographeum, 1982).

Rose, Paul Lawrence, *The Italian Renaissance of Mathematics: Studies on Humanists and Mathematicians from Petrarch to Galileo* (Geneva: Librarie Droz, 1975).

Rosen, Edward, *Three Imperial Mathematicians: Kepler Trapped Between Tycho Brahe and Ursus* (New York: Abaris Books, 1986).

Rossi, Paolo, *Francis Bacon* (London: Routledge & Megan Paul, 1968).

Tournoy, G., J. De Landtsheer and J. Papy (eds.), *Justus Lipsius Europae Lumen et Columen* (Leuven, 1999).

Voelkel, James R., *Johannes Kepler and the New Astronomy* (Oxford: Oxford University Press, 1999).

Voet, Leon, *Antwerp in the Golden Age* (Antwerp: Mercatorfonds, 1973).

Wilson, Edward O., *Consilience, The Unity of Knowledge* (London: Abacus, 1998).

Vöhringer, Christian, *Masters of Netherlandish Art: Pieter Bruegel* (Cologne: Könemann Verlagsgesellschaft, 1999).

Wooding, Lucy, *Tudor England: A History* (London: Yale University Press, 2022).

Wormald, Francis, and Cyril Ernest Wright: *The English Library Before 1700: Studies in its History* (London: University of London, 1958).

Zeeberg, Peter, *Urania Titani, Et digt om Sophie Brahe* (Copenhagen: Museum Tusculanums Forlag, 1994).

Zinner, E., *Regiomontanus: His Life and Work*, trans. E. Brown (Oxford: North-Holland, 1990).

*
* *
*

Articles

Algazi, Gadi, 'Scholars in Households: Reconfiguring the Learned Habits, 1400–1600,' *Science in Context*, 16 (2003).

Almási, Gábor, 'Astrology in the Crossfire: The Stormy Debate after the Comet of 1577', *Annals of Science*, 79 (2022).

Batho, G. R., 'The Library of the "Wizard" Earl: Henry Percy Ninth Earl of Northumberland (1564–1632)', *The Library*, 5th series, Vol. 15 (1960).

Blair, Ann, 'Tycho Brahe's Critique of Copernicus and the Copernican System,' *Journal of the History of Ideas*, 51 (1990).

Daston, Lorraine, 'The Ideal and the Reality of the Republic of Letters in the Enlightenment,' *Science in Context*, 4 (1991).

Dickens, A.G., and Robert Parkyn, 'Robert Parkyn's Narritive of the Reformation', *English Historical Review*, Vol. 62, No. 242 (January 1947).

Hannaway, Owen, 'Laboratory Design and the Aim of Science: Andreas Libavius versus Tycho Brahe,' *Isis*, 77 (1986).

Harkness, Deborah, 'Managing an Experimental Household: The Dees of Mortlake and the Practice of Natural Philosophy', *Isis*, 88 (1997).

Johnstone, Stephen, 'Mathematical Practitioners and Instruments in Elizabethan England,' *Annals of Science*, 48 (1991).

King, D. A., 'Taqi al-Din', *The Encyclopaedia of Islam*, Vol. X, eds. P. J. Bearman, T. H. Bianquis, C. E. Bosworth, E. van Donzel and W. P. Heinrichs (Leiden: Brill, 2000), 132.

Madison, F. R., 'Early Astronomical and Mathematical Instruments. A Brief Survey of Sources and Modern Studies', *History of Science*, 2 (1963).

Moran, Bruce T., 'German Prince-Practitioners: Aspects in the Development of Courtly Science, Technology, and Procedures in the Renaissance,' *Technology and Culture*, 22 (1981).

Pumfrey, Stephen, and Frances Dawbarn, 'Science and Patronage in England, 1570–1625: A Preliminary Study', *History of Science*, 42 (2004).

Shackelford, Jole, 'Tycho Brahe, Laboratory Design and the Aim of Science: Reading Plans in Context,' *Isis*, 84 (1993).

Shapin, Steven, 'Here and Everywhere: Sociology of Scientific Knowledge,' *Annual Review of Sociology*, 21 (1995).

Smith, Jeffrey Cripps, 'Netherlandish Artists and Art in Renaissance Nuremberg', *Simiolus*, 20 (1990–91).

Thomas, Charles B., 'Magic and Mathematics at the Court of Rudolph II', *Elemente Der Mathematik*, Vol. 50, No. 4 (1995).

Wesley, Walther G., 'The Accuracy of Tycho Brahe's Instruments,' *Journal for the History of Astronomy*, 9 (1978).

Westman, Robert S., 'The Astronomer's Role in the Sixteenth Century: A Preliminary Study,' *History of Science*, 18 (1980).

———— 'The Melanchthon Circle, Rheticus and the Wittenberg Interpretation of the Copernican Theory,' *Isis*, 66 (1975).

NOTES

Prologue: Before

1. Grafton, 'Libraries and Lecture Halls,' in Park and Daston (eds), *Cambridge History of Science Vol. 3*, p. 240.
2. Ibid., Park and Daston (eds), 'Introduction: The Age of the New,' p. 2.
3. Grafton, *Defenders of the Text*, p. 145.
4. Augustine, *The City of God* (*De Civitate Dei*), Book V, p. 146.
5. Newman & Grafton, *Secrets of Nature*, p. 15.
6. Singer (ed.), *The Life of Sir Thomas More*, p. 12.
7. Harkness, 'Managing an Experimental Household,' p. 249.

1: Nuremberg

1. Hartmann Schedel, *Liber Chronicarum*, or 'Nuremberg Chronicle' (Nuremberg: Anton Koberger, 12 July 1493). Available online at: https://cudl.lib.cam.ac.uk/view/PR-INC-00000-A-00007-00002-00888/1.
2. Sachs and Amman, *The Book of Trades*, p. 125.
3. Ibid., p. 21.
4. Letter from Regiomontanus to Christian Roder, July 1471, quoted in Rosen, 'Johannes Regiomontanus' in Gillispie & Holmes, *Dictionary of Scientific Biography*, p. 351.
5. Zinner, *Regiomontanus, His Life and Work*, p. 86.
6. Moseley, Adam, 'Regiomontanus Tradelist', a quick and dirty translation. p. 1. Available online at: https://www.academia.edu/28760267/Regiomontanus_Tradelist_pdf.

7. Ibid., p. 3.
8. Ibid., p. 4.
9. Ibid., p. 4.
10. Risk, *Erhard Ratdolt, Master Printer*, p. 34.
11. Zinner, p. 157. Scherp may well have helped fund Regiomontanus' project in some way.
12. Zinner, p. 158.
13. Rose, *The Italian Renaissance of Mathematics*, p. 109.
14. Zaunbauer, 'Albrecht Dürer: A Biography,' in Metzger (ed.), *Albrecht Dürer*, p. 13.
15. Panofsky, *The Life and Art of Albrecht Dürer*, p. 9.
16. Moryson, *An Itinerary*, p. 19.
17. Häberlein, *The Fuggers of Augsburg*, p. 149.
18. Ibid., p. 59.
19. Ibid., p. 53.
20. Smith, *The Body of the Artisan*, p. 76.

2: Louvain

1. The *Compendius Rehearsall* in James Crossley (ed.), *Dr John Dee: Autobiographical Tracts*, p. 5. (Hereafter referred to as *CR*.)
2. Febvre and Martin, *The Coming of the Book*, p. 125.
3. Müller, *The Complete Paintings of Bruegel*, p. 16.
4. Vesalius, *De humani corporis fabrica*, p. 161.
5. Letter from Gemma Frisius to John Dantiscus, 12 December 1539, *CORPUS of Ioannes Dantiscus' Texts & Correspondence* (ed.) Anna Skolimowska (director of the project) and Magdalena Turska, with collaboration of Katarzyna Jasińska-Zdun, dantiscus.al.uw.edu.pl, first published 2010-07-01. Accessed online via: http://dantiscus.al.uw.edu.pl.
6. van Cleempoel, *Scientific Instruments in the Sixteenth Century*, p. 60.
7. Van Ortroy, *Bio-bibliographie de Gemma Frisius*, p. 171.
8. Osley, *Mercator*, p. 185.
9. Ibid., p. 186.
10. See MacTutor's entry on Gerardus Mercator: https://mathshistory.st-andrews.ac.uk/Biographies/Mercator_Gerardus/.
11. van Cleempoel, p. 60.
12. Uppsala Universitetsbibliotek, Carolina Rediviva, H. 154, f. 70.

13. Letter from Gemma Frisius to Johannes Dantiscus, *CORPUS of Ioannes Dantiscus'*, accessed online via: http://dantiscus.al.uw.edu.pl.
14. Crane, *Mercator: The Man who Mapped the Planet*, p. 66.
15. Pettegree, *Europe in the Sixteenth Century*, p. 172.
16. van Cleempoel, p. 61.
17. Shumaker and Heilbron (eds.), *John Dee on Astronomy*, p. 111.
18. Ibid., p. 113.
19. *CR*, pp. 6–7.
20. Crane, p. 145.

3: Mortlake

1. *CR*, p. vii.
2. Ashmole's Preface to British Library, MS Sloane 3188 2r–3r.
3. Several of Dee's books and treatises were written on the request of the queen and her councillors, for example, 'Her Maiesties Title Royall, to many forrain Cuntries, kingdomes, and prouinces...at her Maiesties commandement – anno – 1578.' John Dee, 'Course of the Philosophicall Studies,' London, 1599. Reissued as *The English Experience, its record in early printed books*, published in facsimile, Number 502 (Amsterdam: Theatrum Orbis Terrarum Ltd, 1973).
4. Barnard, McKenzie and Bell (eds.), *Cambridge History of the Book in Britain, Vol. IV*, p. 1. (Hereafter *CHBB IV*.)
5. *CR*, p. 5.
6. Taylor, *Tudor Geography*, p. 269.
7. Dickens and Parkyn, 'Robert Parkyn's Narritive of the Reformation,' p. 66. I am indebted to Dr Lucy Wooding for this reference.
8. Parry, *The Arch-Conjurer of England, John Dee*, p. 32.
9. Calendar of State Papers-Domestic, 1547–1580, 67 (PRO-SP 11/5/34) 8 June 1555, Calais, Thomas Martyn to Edward Courtenay, Earl of Devon.
10. Wormald and Wright, *The English Library Before 1700*, p. 149.
11. Letter to Lord Burghley, 3 October 1574, MS Lansdowne 19, fols. 81–84 in Halliwell (ed.), *A Collection of Letters*, p. 17.
12. As she was described by the Spanish ambassador Count Feria in a letter to Philip II, 14 November 1558, quoted in Loades, *Mary Tudor*, pp. 200–202.
13. Batho, 'The Library of the "Wizard" Earl,' pp. 246–261.

14. Ibid., p. 256.
15. *CHBB IV*, p. 155.
16. Letter to Sir William Cecil, 16 February 1563, written from Antwerp, Calendar of State Papers Domestic XXVII, no. 63.
17. CR, p. 40.
18. CR, p. 30.
19. Bodleian Library MSS Ashmole 487 & 488 *Dee's Diary* published in Fenton (ed.), *The Diaries of John Dee*, p. 44. (Hereafter *Diaries*.)
20. Harkness, 'Managing an Experimental Household,' p. 249.
21. *Diaries*, p. 24.
22. Ibid., p. 94.
23. Lowry, *The World of Altus Manutius*, p. 165.
24. Trinity College Library, Cambridge, MS Adv. d. 1. 26.
25. Tournoy and Papy (eds.), *Justus Lipsius Europae Lumen et Columen*, pp. 35, 42.

4: Kassel

1. Martin Luther, Preface to Philip Melanchthon, *Commentary on the Colossians* (1529), quoted in Otto Kirn, 'Melanchthon', in Samuel Macauley Jackson (ed.), *New Schaff-Herzog Encyclopedia of Religious Knowledge*, Vol. 7, (New York: Funk & Wagnalls, 1908–1914) p.282.
2. Hammer, 'Melanchthon, Inspirer of the Study of Astronomy,' p. 308.
3. Ibid., p. 313.
4. https://orka.bibliothek.uni-kassel.de/viewer/thumbs/1378193947325/9/.
5. Peter Ramus, *Scholarum mathematicarum* quoted in Moran, 'German Prince-Practitioners,' p. 270.
6. Moran, 'Science at the Court of Hesse-Kassel,' p. 91. (Hereafter *Thesis*).
7. Moran, 'Wilhelm IV of Hesse-Kassel,' pp. 72–73.
8. Heidelberg University Library, Manuscript Cod. Pal. germ. 703.
9. Köttelwesch, *Geliebte, Gemahlinnen und Mätressen*, pp. 29–38.
10. Moran, *Thesis*, p. 163.
11. Ibid., p. 164.
12. Johannes Kepler, *De stella tertii honoris in Cygno Narratio Astronomica* (Prague and Frankfurt: Pavel Sessius and Wolfgang Richter, 1601).

13. Moran, *Thesis*, pp. 165–166.
14. Ibid., p. 168.
15. Ibid., p. 168.
16. Moran, 'German Prince-Practitioners,' p. 264.
17. Moran, 'Wilhelm IV of Hesse-Kassel,' p. 82.
18. Moran, 'German Prince-Practitioners,' p. 267.
19. Gingerich and Westman, *The Wittich Connection*, p. 13.
20. Moran, *Thesis*, 210. Letter from Wilhelm to Brahe.
21. Ibid., p. 338.
22. Ibid., p. 340.
23. Moran, *Patronage and Institutions*, footnote 4, p. 137.

5: Hven

1. Thoren, *The Lord of Uraniborg*, p. 14.
2. Christianson, *Tycho Brahe and the Measure of the Heavens*, p. 23.
3. Ibid., p. 25.
4. Thoren, p. 18, footnote 27.
5. Ibid., p. 28. Brahe, *Instruments*, p. 119.
6. Christianson, *Tycho Brahe*, p. 53.
7. Ibid., *Tycho Brahe*, p. 73.
8. Christianson, *On Tycho's Island*, p. 22.
9. Ibid., p. 32.
10. Christianson, *Tycho Brahe*, p. 91.
11. Almási, 'Astrology in the Crossfire', pp. 9–10.
12. Ibid., p. 9.
13. Letter from Brahe to Schultz printed in Thoren, p. 268.
14. Christianson, *On Tycho's Island*, pp. 89–90.
15. Gingerich and Westman, *The Wittich Connection*, p. 18.

6: Prague

1. Bergeron's *Description of Prague in the time of Rudolph* II, in Bažant et al. (eds.), *The Czech Reader*, pp. 77–78.
2. Moryson, *An Itinerary*, p. 14.
3. Evans, *Rudolph II and His World*, p. 122.
4. Ibid., p. 44.
5. Ibid., p. 45.

6. Bergeron, *Description*, p. 77.
7. Findlen, 'Cabinets, Collecting and Natural Philosophy' in Fučíková, *Rudolf II and Prague: The Court and the City*, p. 216.
8. Ibid., p. 216.
9. Read more about the museum on their website: http://www.alchemiae.cz/en
10. *Diaries*, p. 134.
11. Ibid., p. 140.
12. Ibid., p. 143.
13. Ibid. pp. 205–206.
14. Ibid., pp. 174–175.
15. Ibid., p. 151.
16. Ibid., p. 233.
17. Ibid., p. 204.
18. Ibid., p. 238.
19. Bergeron, *Description*, p. 78.
20. Tycho Brahe to Rosenkrantz, 30 August 1599 [old style], Thoren, pp. 412–413.
21. Tycho Brahe to Sophie Brahe (Lange), 21 March 1600, Thoren, pp. 508–509.
22. Thoren, p. 433.
23. Gingerich, 'Johannes Kepler' in *Dictionary of Scientific Biography*, p. 289.
24. Osterhage, *Kepler: The Order of Things*, p. 39.
25. Voelkel, *Johannes Kepler and the New Astronomy*, p. 57.
26. Connor, *Kepler's Witch*, p. 252.
27. Fučíková, *Rudolf*, p. 63.
28. Gingerich, 'Johannes Kepler' in *Dictionary of Scientific Biography*, p. 299.
29. Ibid., p. 290.

7: Atlantis

1. Vickers (ed.), *Francis Bacon, The Major Works*, p. 478.
2. Jardine and Silverthorne (eds.), *Francis Bacon: The New Organon*; New Organon I, *Aphorism*, (Cambridge: Cambridge University Press, 2000) p. 129.

3. Thomas Fuller, quoted in Jardine and Stewart, *Hostage to Fortune*, p. 25.
4. Spring, *Mistress of Gorhambury*, p. 9.
5. Ibid., p. 9.
6. Allen, *The Cooke Sisters*, p. 124.
7. Jardine and Stewart, *Hostage*, p. 27.
8. Ibid., p. 53.
9. Ibid., p. 49.
10. Vickers, p. 471
11. Jardine and Stewart, p. 63.
12. Allen, p. 144.
13. Spedding, Ellis and Heath, *Collected Works of Francis Bacon*, Vol. I (London: Routledge, 1996), p. 7.
14. *The Advancement of Learning Book II* in Vickers, p. 172.
15. Henry, *Knowledge is Power*, p. 43.
16. MS Harley 541 British Library, 35. View online at: https://upload.wikimedia.org/wikipedia/commons/4/49/Gesta_Grayorum._1688_%28IA_gestagrayorum16800grayrich%29.pdf.
17. *Diaries*, p. 46.
18. Ibid., p. 2.
19. Spring, p. 32.
20. Constantijn Huygens, quoted in Smith, *The Body of the Artisan*, p. 163.
21. Letter from Sir John Holles to Lord Gray 27 February 1613 in Strong, *Henry Prince of Wales and England's Lost Renaissance*, p. 2.
22. Andreä, *Christianopolis*, p. 160. Available online: https://archive.org/details/christianopolis00andr/page/228/mode/2up?view=theater.

Epilogue: Beyond

1. Moran, 'Courts and Academies,' in Park and Daston (eds.), *Cambridge History of Science, Vol. 3*, p. 270.
2. Almási, 'Astrology in the Crossfire,' p. 26.

INDEX